Acting as a Way of Salvation

A STUDY OF RĀGĀNUGĀ BHAKTI SĀDHANA

David L. Haberman

New York Oxford
OXFORD UNIVERSITY PRESS
1988

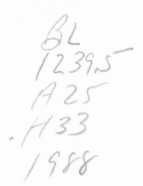

Oxford University Press

Oxford New York Toronto
Delhi Bombay Calcutta Madras Karachi
Petaling Jaya Singapore Hong Kong Tokyo
Nairobi Dar es Salaam Cape Town
Melbourne Auckland

and associated companies in
Berlin Ibadan

Published by Oxford University Press, Inc.,
200 Madison Avenue, New York, New York 10016

Library of Congress Cataloging-in-Publication Data

Haberman, David L., 1952–
Acting as a way of salvation: a study of rāgānugā bhakti sādhana
David L. Haberman.
p. cm.
Bibliography: p.
Includes index.
ISBN 0-19-505321-4
1. Acting—Religious aspects—Hinduism. 2. Rupāgosvāmī, 16th
cent. 3. Bhakti. 4. Hinduism—Doctrines. I. Title.
BL1239.5.A25H33 1988
294.5′22—dc 19 87-33960
CIP

9 8 7 6 5 4 3 2 1
Printed in the United States of America
on acid-free paper

To all those who
helped me glimpse
some of the mystery of
Kṛṣṇa's Līlā

Foreword

Oscar Wilde once wrote Ada Leverson, asking her about Max Beerbohm: "When you are alone with him, does he take off his face and reveal his mask?"* Wilde was obsessed with, among other things, masks, for there is truth in the observation that only when a mask is being worn does one express the truth, and only then can the true self be discerned. The problem, of course, is that when one has worn a mask long enough it is sometimes no longer possible to tell which is mask and which is face.

It is a mark of the creativity of the Vaiṣṇavas of Bengal (or the Gauḍīya Vaiṣṇavas, as David Haberman chooses to call them here) that such an interesting question can form the cornerstone of a religious and philosophical system, for definitions of the self, or selves, and the roles each of these play are critical to Vaiṣṇava considerations. The script for the roles is Reality, and the playwright divine. In the Vaiṣṇava system, the Stanislavski method is intensified to the *n*th degree: the part for which one trains is, ultimately, one's true self, and the transformation, when it comes about, lasts for longer than the play runs. It is complete, and final, and forever.

Haberman points out to us that the English word "play" has several meanings. It signifies "game" or "drama," both of which are segments of action sometimes only metaphorically related to reality, defined and structured in such a way as to be made comprehensible in abbreviated time and space. But the Sanskrit word *līlā,* which is also usually translated "play," has an additional connotation, for it suggests the vast and unknowable mind of God, only tiny bits of which can be understood by our impoverished human processes. As a game imposes rules on random behavior, or as a drama editorializes upon segments of human experience, so the reality of human life is a definable fragment of the Real. The relationship is not metaphor but metonymy: by participation in the real one participates also in the Real. The trick is to understand that. And since one's small mind is not capable, one enters the play of God, the *līlā,* by means of drama. One understands a small part of the mind of

*Quoted in Richard Ellman's biography *Oscar Wilde* (New York: Knopf, 1988), p. 309.

God by directed experience, by playing one's role on what is, ultimately, the divine stage.

Nor are there auditions for the parts. Everyone has a role, and it is self-selected. The play was written long ago, before time began. It is the play of Kṛṣṇa, made known on earth through the text called the *Bhāgavata Purāṇa,* and it is infinite. There are roles for all who choose to be devoted, who are willing to train themselves until they understand that they are, in fact, in the world of the Real, the friends or parents or servants or—most significantly—the lovers, of God. Because the drama is divine, the stage is eternity, the time frame is no longer act and scene. The real world and the Real world are revealed to be the same.

If this seems somewhat esoteric, it is because one of the cornerstones of the Vaiṣṇava system—which, as Haberman points out, is a system of both thought and method—is the concept *bhedābheda,* the paradoxical and simultaneous immanence and transcendance of the divine. The concept is further described as *acintya,* not to be understood rationally. If there is a distinctive feature of Indian mysticism, it is that it is practical and experiential: it not only perceives the condition of the ultimate relationship of God to man, it also tells us just how to get there. The detailing of this process is one of the fascinating points of this book.

The concept and the process, and therefore the book, are both specific and *not* specific to India. There are of course many religions in which ecstatic experience is valued and is the ideal, many in which a model, or a paradigm of the process, is presented for emulation. But there are few, if any, that are so elaborate, so detailed, and so entwined with esthetic understanding that are based on the idea that there is a realm of the spirit which God shares with man. The imitation of Christ is not an unfamiliar theme in Christianity but, in itself, I don't think it aims one toward identity. The systems may be similar, in that in Christianity one does indeed participate in the Body of Christ through an institution, the Church, but Christ is unique in being God as well as man. His deeds, because of his divine nature, are by definition perfect. One can imitate them, for they were after all performed by a man, but through imitation one does not become Christ. The Vaiṣṇava both becomes and does not become Kṛṣṇa, for the stage on which the drama is being acted and the dramatis personnae are extensions of him.

So this is the context of David Haberman's concern, and an interesting and significant concern it is. In no other way of thinking, to my knowledge, is there such an intriguing and precise analysis of the relationship between esthetic and religious experience, of the proposition that the yogin, abstracted from the world in his meditation, is like the reader of a poem, totally taken up into the poetic world and oblivious to all else around. It sheds a different light, I think, on the relationship between the finite and the infinite. The book is a meditation on man's mental and spiritual access to worlds in which time and space have different meanings. It goes far beyond sectarian Hinduism and says things about our humanity that I, at least, did not know before.

Edward C. Dimock, Jr.
University of Chicago

Acknowledgments

The staging of this book has received the generous help of many directors, coaches, fellow actors, and stage hands. I cannot possibly mention them all, but I would like to offer a special thanks to the following.

This book began as a doctoral dissertation at the University of Chicago. The members of my committee provided much advice and assistance. Edwin Gerow initiated me into the joys of the Sanskrit language and shared with me his astute knowledge of the Indian *rasa*-theory. He also took the time to read portions of the *Bhaktirasāmṛtasindhu* with me. Throughout the project, Frank E. Reynolds encouraged me to conceive of the particular study of Rāgānugā Bhakti Sādhana in a context pertinent to the larger issues of the history of religions. I owe special thanks to Edward C. Dimock, Jr. who introduced me to the works of the Vṛndāvana Gosvāmins, particularly the *Bhaktirasāmṛtasindhu* of Rūpa Gosvāmin, and has inspired me in many ways. He also graciously agreed to write the foreword to this book. I am especially grateful to Wendy Doniger O'Flaherty, the chief director of my dissertation. She patiently read each chapter as it was produced, offering relevant comments that demonstrated a deep understanding of the issues I struggled with. Throughout my graduate studies, she gave the kind of support that makes her a paradigmatic advisor.

This book is also the result of the generous aid of many in India. In particular, I want to thank Shrivatsa Goswami of the Śrī Caitanya Prema Sansthāna in Vṛndāvana who helped transform Vṛndāvana from a strange place into a home and introduced me to the community there. His friendship is one of the valuable treasures I discovered in India. I owe special thanks to Acyuta Lal Bhaṭṭa, resident of Vṛndāvana and lecturer at the Champa Agrawal Inter College in Mathurā, who graciously took time out of his busy schedule to read with me Rūpa's *Bhaktirasāmṛtasindhu.* The delightful hours we passed together will always flavor my interpretation of this vast text. Dr. Bhaṭṭa also read with me the *Rāgānugāvivṛtti* of Rūpa Kavirāja and shared with me his ideas on what proved to be a very difficult text. To the staff of the Vrindaban Research Institute, I am grateful for their assistance with the manuscript research I conducted. And finally, my thanks goes out to the *sādhaka*s of

Vṛndāvana and Rādhākuṇḍa and to Asim Kṛṣṇadāsa for introducing me to some of the true mystery of Vraja.

Friends and colleagues have been an invaluable aid throughout this project. Neal Delmonico took the time to read Rūpa Kavirāja's *Rāgānugāvivṛtti* with me in Chicago and offered his keen insights regarding a study of Gauḍīya Vaiṣṇava *sādhana.* Tony K. Stewart and Robert D. Evans shared with me the fruits of their ambitious bibliographical work on Gauḍīya Vaiṣṇavism. Veena Das and John S. Hawley read earlier drafts of this book and offered helpful suggestions. Elizabeth A. Isacke contributed significantly in a number of ways. Many thanks to all. I am especially grateful to Sandra H. Ducey for her valuable editorial assistance and sustained emotional support.

Preliminary versions of small portions of this book have appeared as articles in the *Journal of the American Academy of Religion* ("Imitating the Masters: Problems in Incongruity," March 1985: 41–49); in *South Asia Research* ("Entering the Cosmic Drama: Līlā-Smaraṇa Meditation and the Perfected Body," May 1985: 49–57); and as a chapter, "The Religious Esthetics of the Bengali Vaiṣṇava Community at Rādhākuṇḍa" (pp. 47–51), in *Bengal Vaiṣṇavism, Orientalism, Society and the Arts,* edited by Joseph T. O'Connell (East Lancing: Michigan State University, 1985).

The Department of Education made my year of research in India possible with a Fulbright-Hays Fellowship. The Committee on Southern Asia Studies and the Divinity School Institute for the Advanced Study of Religion, both of the University of Chicago, provided the funds for much of the writing of this book. I am grateful to all.

Williamstown, Massachusetts D.L.H.
January 1988

Contents

Note on Translations and Transliteration

The translations of Sanskrit, Bengali, and Hindi which follow are all my own unless otherwise attributed. I have tried to follow the conventions of contemporary Sanskritists even when transliterating terms from Bengali and Hindi sources, for the purpose of consistency. Thus, for example, although one may more accurately transliterate Vṛndāvana as "Brindābon," from Bengali sources, and as "Vrindāban" from Hindi sources, I have transliterated it as "Vṛndāvana" in all cases, unless it was spelled otherwise in an English text.

Diacritics have been employed to guide pronunciation; again, Sanskrit conventions have been followed. The long vowels, *ā, ī,* and *ū* are marked with a length mark (-) and are pronounced approximately the same way as the corresponding vowels in the English words f*a*ther, m*ee*t, and p*oo*l. The vowels *e* and *o* are always long and pronounced like the vowels in pr*e*y and m*o*w. The diphthongs *ai* and *au* are pronounced approximately like the vowel sounds in the words m*a*ss and c*au*ght. The short vowels *a, i,* and *u* are pronounced like the vowels in the words b*u*t, s*i*t, and p*u*ll. Vocalic *ṛ* is also short and is pronounced like the *ri* in "ring."

Consonants are usually pronounced as in English, with the following qualifications: The *c* is pronounced as the *ch* in "church," the *j* as the *j* in "jungle," while *ś* and *ṣ* are both pronounced like the English *sh.* The aspirated consonants should be pronounced distinctly; for example, *gh* as in "doghouse," *th* as in "boathouse," and *bh* as in "clubhouse." Cerebral consonants are marked with a dot under the consonant (e.g., *ṭ* and *ṇ*) and are pronounced with the tip of the tongue retroflexed to strike the roof of the mouth. The *ṇ* is pronounced like the *n* in "jungle" and the *ñ* is pronounced like the French palatalized *n* (written *gn*). The visarga (*ḥ*) is a light voiceless aspirate. A glossary of frequently used Sanskrit terms is provided at the end of this book.

ACTING AS A WAY OF SALVATION

1

Introduction

I hold the world but as the world, Gratiano—
A stage, where every man must play a part.

WILLIAM SHAKESPEARE, *The Merchant of Venice*

There are as many realities as you care to imagine.

LAWRENCE DURRELL, *Balthazar*

This is a study of a religio-dramatic technique known as Rāgānugā Bhakti Sādhana, a technique that can tell us much about how religious experience is constituted. The study is a result of the union of two interests of mine. I have long been fascinated with the process of entering a religious reality; the *how* of religion has intrigued me, puzzled me. The great masters of religious studies have been quite successful at stirring my interest and imagination with descriptions of the fabulous realities of the world's religious traditions, but have often done so in a way that made these realities seem inaccessible, far removed from human actors. A question remained for me. How does a person come to inhabit a religious reality?

My second interest is the theatre. I have a longstanding fascination with the experience that theatre can evoke, its ability to transport one into another world. These two interests converged the day I encountered the works of Rūpa Gosvāmin. One of the chief theologians of a Hindu devotional (*bhakti*) movement, which has its roots in sixteenth-century Bengal and is known as Gauḍīya Vaiṣṇavism,[1] Rūpa systematized a technique of shifting from one reality to another; this technique was heavily dependent on dramatic experience.

This book explores the religio-dramatic technique that Rūpa established as a means for entering and participating in ultimate reality as envisioned by the Gauḍīya Vaiṣṇavas. Very briefly, the Rāgānugā Bhakti Sādhana is a technique informed by classical Indian aesthetic theory and involves assuming the role of a character in the ultimate reality. Dramatic technique is especially appropriate here because the Gauḍīya Vaiṣṇavas conceive of ultimate reality as a cosmic drama, the eternal play of Kṛṣṇa. The Gauḍīya Vaiṣṇavas claim that there is a whole world of which we are normally unaware, and that each of us has a "double" in that world. The goal of Rāgānugā Bhakti Sādhana is to shift identity to that "double," called by this tradition the "perfected form" (*siddha-rūpa*), which is one's true and ultimate identity. Salvation, to the Gauḍīya Vaiṣṇavas, is unending participation in the cosmic drama, and the skills of the actor are employed in pursuit of the true identity which allows such participation. Because of my peculiar interests and academic background I will obviously be taking a very different approach to a study of the Rāgānugā Bhakti Sādhana than would a Gauḍīya Vaiṣṇava. To give the reader some indication of my approach, I begin with a brief preliminary discussion of the nature of the process by which any reality is entered, why this process poses a religious problem, and how I think a religio-dramatic role-taking technique attempts to solve it.

An examination of the way in which reality is entered requires an introductory consideration of the nature of reality.[2] The contention that reality, for human beings, is neither fixed or singular is now commonly accepted in academic circles. Though we possess a similar biological body, multiple realities or worlds of meaning are available to us. Much ethnological evidence supports this claim; one has only to survey the diversity of cultures to be struck by the enormous variability in the ways of being human, in people's sense of what is real and meaningful. Reality is culturally constructed. It is not a given for the human being; if it were, the entire process of *entering* a reality would be superfluous. But this is not the case. Human beings are not fixed psychologically, but instead manifest an immense plasticity in the arena of reality construction. Our malleable nature enables us to experience a wide and seemingly endless range of possible realities.

This facet of human experience, sometimes called "world-openness,"[3] has been well noted by many working in the human sciences.[4] More than any other single factor, the undeveloped nature of our instincts seems to account for our "world-openness," an ability to occupy a plurality of worlds. Our biological condition is typically characterized by the lack of instincts, which provide programmed, stable direction for the conduct of other animals. Therefore, any degree of stability that we are to achieve must be provided by nonbiological means. We must construct our own world. And indeed, human beings are amazingly successful at this miraculous task, for when we examine the many cultures of the world we do not find people in a state of constant and unstable flux. Rather, they are located within a definite reality, complete with structures of meaning and patterns of conduct. We must then ask: What guides human conduct? By what means do humans come to inhabit a particular reality in this vast sea of seemingly infinite possibilities?[5]

"Habitualization" provides the behavioral direction lacking in the human biological makeup. Habitualization, or routinized activity, saves the individual from the pressures of facing a multitude of choices at each moment of life; the terrors of infinite possibility are thus minimized. The individual can thereby act in what appears to be a sure manner. Within the social setting habitualization eventually gives way to institutionalization. An institution is habitualized action, shared by a group, that renders behavior typical and predictable. Institutions guide and control behavior by setting up predefined patterns that funnel activity in one direction, thereby delivering the human being from the chaotic realm of boundless choice. The institution, then, acts in lieu of an insufficient instinctual nature.[6]

Moreover, institutions are what lead individuals into a particular world of meaning, for they are the objectified activity by which individuals grasp the meaning of those around them. Institutions are objectified forms of subjective experience. They define and delineate every aspect of the social world. Individuals born into a particular society meet the institutional structures as preexistent, a "given" for the entire community. To enter and participate in a community's world of meaning, individuals must incorporate these institutions as have those before them. This is the ongoing process of socialization.

Socialization is the process by which the social reality is transmitted to the individual through the institutions of that society. How does this occur? A child is not born into a vacuum, but rather into a world already peopled, most particularly by those directly caring for the child, whom we might call "significant others."[7] Through their daily actions these significant others present the shared social world to the child. Institutional order or social reality, then, is not presented in an abstract manner, but through the objectified activity of *individuals*. Hence distinct individuals serve to define reality. The objectified activities of these individuals are called "roles." Roles are behavioral guides, patterns of attitudes and actions which channel the conduct of an individual.[8] The construction of roles is a necessary correlate of the institutionalization of conduct, for roles represent the institutional order for the individual. Therefore, all social conduct involves roles; these are the means by which the individual participates in the social world of meaning. The significant others are thus the child's guides into social reality.

The process of socialization takes on a dramatic quality, for a great deal of role assimilation depends on the direct observation of the person enacting the role that the child is destined to assume. These significant others embody the roles and continually actualize the social drama on the stage of the child's home environment. The child, however, does not begin to enter and participate in the social drama until she or he begins to take part in that drama herself or himself. This is accomplished by internalizing the role presented by a significant other. Identity is not genetically determined; it must be generated in the process of socialization. By internalizing the role of significant others and making them her or his own, the child develops an identity.

Identity, then, is the vehicle into the world of the social reality. When a role is internalized as an identity, the world of that role becomes subjectively real. Reality is dependent on one's perspective or location, and identity is defined as one's place, one's location in a certain world.[9] Entering a reality, then, is a locative process. Since reality depends on location, and identity is defined as location in a certain world, reality and identity are inextricably linked. In fact, identity—an internalized role located within a particular reality—is the very vehicle into that world of meaning. Thus the analysis of roles is of particular importance because it reveals the way in which a reality is entered and made subjectively real for individuals. If we transplant this role theory from its home environment of sociology into the arena of religious experience, we then have a method with which to approach the issue of entering a religious reality.[10]

The recognition that we can experience a multiplicity of realities poses a serious religious problem. The fact that ways of being and avenues of possibility are practically infinite is viewed as potentially dangerous. The human being as *homo religiosus* is not satisfied to occupy just any reality, but instead thirsts for the "Real."[11] The religious person seeks to participate fully in a reality that stands qualitatively above all others.[12] He or she strives to perform not just any acts, but acts that place him or her in harmony with the paramount reality. To act in accordance with what is considered the "ultimate reality," a reality which exists against the backdrop of a variety of ways of being, the religious individual also has need of guides to channel conduct in a manner conducive to what is believed to be "Real." Such guides, we have argued, are embodied in "institutions." Thus comes the realization that institutions are not the enemy of religious persons, but rather their natural habitat.

Many have noted that the human being is capable of shifting from one reality to another. However, a number of writers—such as Alfred

Schutz, who resorts to a language of "shock" to explain religious experience[13]—contend that there is no formula for this transformation. A religious community concerned with reaching the other reality, however, cannot depend on haphazard shocks. It must devise and maintain systematic techniques that will enable its practitioners to attain a position in what is defined as the ultimate reality. In the Hindu tradition such a technique is called *sādhana* ("a method of realization").

Because socialization is the process that locates one in a particular reality, what is in effect required is a religious re-socialization—or *conversion,* to use the more familiar psychological term. This latter process relocates one in a new religious world of meaning. This requisite shift from one world to another does not pertain to all types of religion; in some cases, the social and religious worlds are hardly differentiated. But in a society where even a slight degree of plurality exists, religious groups frequently hold to a reality quite distinct from the dominant social reality. Under these conditions the social reality is devalued and contrasted with the "ultimate" reality. In this case, the ultimate reality is by definition distinct from that of the ordinary world in which one finds oneself.

The claim that there is a more valuable world beyond the socially defined one demands a method by which one can enter the highly esteemed religious world. Considering the important function of institutions in the process of primary socialization, it would seem that similar guiding influences would be necessary for religious communities in quest of an ideal religious world.[14] We would expect to find institutionalized roles significantly employed by these religious communities. A number of such roles for the Gauḍīya Vaiṣṇava tradition will be examined in this study. I disagree with the claim of Schutz and others that there is *never* a formula for the transformation which enables one to pass over into a new reality. A study of Rāgānugā Bhakti Sādhana will demonstrate that there is in fact such a formula, and that it is as follows: One enters the religious reality by assuming, *via role-taking,* an identity located within that reality. The new identity is the vehicle to the new reality. Close attention, then, should be given to role-systems and the ritual structures devised by religious traditions to construct new identities for concerned individuals, thereby transporting them to a new concomitant reality.

Gauḍīya Vaiṣṇavas claim that a new identity is the entrance pass to the highest reality of Kṛṣṇa's play. The social identity is the identity associated with the observable role of the individual in society. Religion may legitimate this social identity, but it rarely equates it with the whole person or accepts it as the individual's ultimate end. Religion frequently provides an alternate identity by opening up a "transcendent identity," situated in a universe of meaning which transcends the determinism of the social identity and social world.[15] Rūpa Gosvāmin and the Gauḍīya Vaiṣṇavas are concerned with realizing the true identity defined as the *siddha-rūpa,* one's eternal part in the cosmic drama. An investigation of the techniques and dynamics of the processes the Gauḍīya Vaiṣṇavas have devised, with which to construct a transcendent identity for their participants and thereby lead them to the ultimate world of Kṛṣṇa's play, will afford us an opportunity to glimpse how religious experience is constituted and a religious reality is entered.

Exemplary individuals are extremely important in the process of entering and participating in any world or reality, social or religious. Since these individuals portray roles that serve as the guides into the new reality, we would necessarily expect to find their presence in any process of reality construction. The exemplary individual—or "significant other"—of primary socialization is one who is physically present to objectify the social world for the child; however, in the religious quest, where the aspirant is attempting to reach an "unseen" reality[16] and a transempirical world is valued over the primary social world, a problem arises. Who fills the essential position of the significant other? Where are the viable role models in religious action which is concerned with a transcendent reality, a reality beyond ordinary perception? I propose the choice of mythic models,[17] exemplary divine figures who themselves reside in the ultimate reality, whom I shall call "paradigmatic individuals."[18]

What distinguishes religious role-taking from other forms of role-taking is the nature of the paradigmatic individual. The paradigmatic individual is a figure who is not physically present, but a divine or superhuman being inhabiting a mythical world. This assertion supports the contention that religion can be differentiated from other culturally constituted institutions only by virtue of its reference to superhuman beings.[19] Scripture, mythical narratives, and sacred biographies, which establish and maintain the exemplary roles of these individuals, and present them as paradigms to the ongoing religious community, serve an extremely important function; in addition to exploring the great puzzles of life, they contain ideal scripts for the religious life. The enactment of the paradigmatic roles found within such scripts is the way I take Mircea Eliade's notion of religion as "living a myth."[20]

The paradigmatic individual is more distinctive than a social role; the paradigmatic individual is a particular mythological character, a condensation of cultural values. The good Hindu wife, for example, is a social role; but, as we shall see, Rādhā is a paradigmatic individual of condensed cultural values, who presents perfection by means of a very particular mythological identity. Moreover, the paradigmatic individual, as exemplified by Rādhā, frequently goes against or disrupts a dominant social role, and as such serves as a vehicle to a reality beyond the social.

How are such figures utilized as vehicles of transformation? Through ritual imitation;[21] culturally established imitative activity transforms identity and situates one in the reality of the paradigmatic individual being imitated. Japanese Shingon monks, for example, ritually enact the role of Mahāvairocana. "In a word, the essence of Kūkai's Esoteric Buddhist meditation is simply 'imitation.' . . . To practice the samadhi of Mahāvairocana is to imitate it through one's total being—physical, mental, moral, intellectual, and emotional—like an actor acting alone on stage."[22] One might also think of how the "imitation of Christ" informs much ritual activity within the Christian traditions.[23] And sincere imitation eventually gives way to becoming. The transformation of identity is one way of defining salvation. In these terms, salvation involves a shift from the socially given identity to a new identity located in a reality believed to be ultimate.[24]

Religious peoples have used many ingenious devices to identify with a particular paradigmatic individual and thereby enter his or her world of meaning. One of the most powerful techniques utilized for transforming identity has been dramatic acting. The religions of the world have produced a wealth of dramatic traditions, and one can readily understand how the concept of the actor's trade—representing a particular role on a given stage—naturally lends itself to the implementation of the transformative process of role-taking. Embodying a paradigmatic identity through dramatic technique produces change; the religious world is attained as this role is enacted.

Masks are frequently used in dramatic performances to aid in the transformative process. A Hopi writes of the ability of the kachina mask to project the wearer into the spiritual world of the kachinas: "I feel that what happens to a man when he is performer is that if he understands the essence of the kachina, when he dons the mask he loses his identity and actually becomes what he is representing."[25] The Gauḍīya Vaiṣṇavas have explored a different means: Rāgānugā Bhakti Sādhana involves dramatic performance on a stage visualized in meditation.

The ability of dramatic role-taking to transform identity, and thus to carry the actor into the world of that role, has been thoroughly explored by Russian director and philosopher Constantin Stanislavski. Stanislavski drew critical attention to the powerful affectivity of dramatic acting, to the ability of dramatic acting to change the life of the actor.[26] He observed that significant changes occur for the actor as he or she identifies with a role and is influenced by its particular way of being. For the true actor, a whole new world is opened up by the role.[27] The result of true acting for Stanislavski is to live the role, to experience the world of the character being enacted. As the fictional director Tortsov, Stanislavski informs a young actor of what happens when the moment of true acting arrives:

> Your head will swim from the excitement of the sudden and complete fusion of your life with your part. It may not last long but while it does last you will be incapable of distinguishing between yourself and the person you are portraying.[28]

Stanislavski called this experience of complete identification with the world of the character "reincarnation."[29] For Stanislavski, the successful actor was the one who could momentarily leave the socially constructed reality and enter into that of the character in the play. His cheer was: "Live your part!" When the actor conforms to the part and achieves "reincarnation," a new identity, a new being is created. "Our type of creativeness (i.e., True Acting) is the conception and birth of a new being—the person in the part."[30] Dramatic acting, Stanislavski confirms, is a powerful means of role-taking, which leads the actor to a new identity and a world concomitant with that identity. When these techniques are applied to an "ultimate" role, the outcome is profound. While the Method actor seeks temporary identification with a part, the religious aspirant seeks permanent identification with a role defined by a paradigmatic individual for the purpose of salvation. The Gauḍīya Vaiṣṇava, we shall see, strives to realize a permanent part in Kṛṣṇa's play.

The following pages explore the Rāgānugā Bhakti Sādhana, a religio-dramatic technique in which the practitioner identifies with paradigmatic individuals from Vaiṣṇava mythology, with the aim of making a permanent entrance onto the ultimate stage of Kṛṣṇa's drama. Since an understanding of this practice depends on some knowledge of Indian aesthetics, I begin in Chapter 2 by examining major developments and issues of Indian dramatic theory, focusing particular attention on religious concerns. I then go on to explore Rūpa Gosvāmin's unique application of dramatic theory to Hindu devotionalism in Chapter 3. To better understand the dynamics of the Rāgānugā Bhakti Sādhana, it will also be necessary to provide, as in Chapter 4, a bit of background information on the Gauḍīya Vaiṣṇava tradition, exploring its historical setting, its need for a powerful transformative technique, and its mythological world filled with paradigmatic individuals. In Chapter 5, I examine the Rāgānugā Bhakti Sādhana itself, particularly as set forth by Rūpa Gosvāmin in his *Bhaktirasāmṛtasindhu,* and the fascinating Gauḍīya Vaiṣṇava version of the "double," the *siddha-rūpa.* In Chapter 6, I investigate some interesting historical developments in the practice, and go on in Chapter 7 to present some contemporary expressions of Rāgānugā Bhakti Sādhana, based on field research conducted in Vraja during 1981–1982. I conclude this work by suggesting what Rūpa's technique may tell us about religious experience in general.

2

Religion and Drama in South Asia

The secret of all art is self-forgetfulness.

ANANDA COOMARASWAMY, *The Mirror of Gesture*

The rather ambiguous attitude toward drama in the West was fairly well established by Plato and Aristotle. These two thinkers agree with Indian theorists of drama in defining drama as a kind of imitation. Plato, however, considered this fact to have damaging consequences; in the *Republic* he claimed that imitative arts lead us farther from the truth.[1] Plato argues that there are three types of people: the philosophers, who know the truth or the Idea; the craftsmen, who copy the Idea and produce appearances; and the imitative artists, who copy the appearances and are thus three steps removed from the truth. The product of the latter group, imitative art, tricks us into confusing appearances for the real thing and thereby leads us farther from the truth. Plato declared that tragedy, or drama, is imitative art of the "highest possible degree" and therefore ranks among the worst of the arts. Because of this, Plato refused to allow drama in his ideal republic.

Aristotle gave a much more positive assessment of drama and, in fact, of imitation in general.[2] Imitation, according to Aristotle, "is innate in men from childhood" and is a valuable source of learning. Aristotle agreed with Plato that drama is a kind of imitation, but argued that it produces a valuable experience, known as *catharsis*. Aristotle also assessed the value of drama in terms of what, or rather who, it imitates. All imitation imitates agents—human beings who can be better than we are, worse than we are, or equal to us. For Aristotle, the worthiness of the agent determines the quality or affective meaning of the imitative art. Tragedy, he declares, is an imitation of the actions of worthy agents, and as such is a valuable source of learning and of catharsis.

Indian critics of drama also speak of drama as imitation (*anukṛti*), but following the authority of the *Nāṭya-śāstra,* they tended unanimously to agree that it had at least a relatively positive value and very often great religious value. The fact that the dramatic experience removes one from ordinary life certainly did not trouble Indian philosophers as it had Plato; in fact, it is precisely this quality of the dramatic experience that gives the imitative art of drama its value to Indian philosophers of aesthetics. The Indian aestheticians agree with Aristotle that drama is a valuable source of learning, but do not define its distinctive value in terms of the agent it imitates. Dramatic experience is valuable for the Indian aestheticians because it is "generalized"[3] experience. To understand the particular value of aesthetic experience for the Indian theorists, we must begin with a close examination of their aesthetic principle, known as *rasa.* A study of the development of the *rasa* theory will also provide us with a solid base upon which to examine how Rūpa Gosvāmin employed and changed the accepted aesthetic theories of his time in his efforts to delineate the process of *bhakti.*

Bharata's *Rasa* Theory

Rasa originally meant "sap," "essence," or "taste." Though it retains this original meaning, in the context of aesthetics it can perhaps best be translated as "dramatic sentiment" or "aesthetic enjoyment." The question of when the specifically aesthetic flavor of *rasa* first appears is debated. Yet I find no reason to doubt Edwin Gerow's assertion: "Taken as a whole, the sketch of *rasa* in the *Nāṭya-śāstra* suggests strongly that the *rasa* developed its first 'aesthetic' overtones in the context of the Sanskrit dramas of the classical period."[4] Therefore, *rasa* as an aesthetic principle should first be understood as a distinctive feature of the dramatic experience; that it occurred first in the context of the drama is a crucial, rather than an incidental, factor in its definition. Indeed, the oldest work to mention *rasa* as a definable aesthetic principle is the *Nāṭya–śātra*[5] of the legendary Bharata. Consequently, an examination of the *rasa* theory of Bharata is the starting point for any adventure into an Indian understanding of dramatic experience.

Bharata explains in the beginning of the *Nāṭya-śāstra* that at the commencement of the Tretā–yuga, the age subsequent to the perfect age, people began to lose their way and so to experience desire, greed, jealousy, and anger. Hindu psychology generally attributes emotions of this sort to the fundamental problem of egoism arising from the perception of oneself as a distinct individual separate from and in competition with other distinct individuals. According to Hindus, this perception is the result of a gradual cosmic disintegration. Hindu cosmology involves a theory of de-evolution. For the sake of creation, the one undifferentiated primal entity became many.[6] As the division increased, egoistic emotions—desire, greed, jealousy, and anger—developed and intensified. It is because of the appearance of these new egoistic emotional problems, Bharata tells us, that people's happiness became mixed with sorrow. To remedy this situation, all the gods, headed by Indra, approached Brahmā, the Supreme Creator, and asked him to create a new Veda which was both audible and visible. Brahmā, accordingly, entered into a yogic state and fashioned, from the four existing Vedas, the fifth Veda—drama (*nāṭya*).

Brahmā then remarked: "This drama will therefore provide instruction to everyone in the world through all its actions, emotional states (*bhāva*s), and *rasa*s."[7] Hence, we see early on, drama was held in South Asia to possess an edifying quality, because of its ability to produce a certain experience called *rasa.* Within the mythological account of the origins of drama as expressed in the *Nāṭya-śāstra,* we glimpse the design to evoke within an audience a particular aesthetic-emotional experience, known as *rasa,* that would cope with or transcend the problems of egoism. From Bharata's time onward, this was to be the central objective of any drama. One can readily understand from this, then, that dramatics developed in India in a very different direction than it did in the West, which for the most part followed Aristotle's assertion that plot was the soul of and organizing principle for drama.[8]

Bharata's central problem in the *Nāṭya-śāstra* was to determine how particular emotions could be evoked in the audience of a drama. This endeavor led him to a very sophisticated analysis of human emotional nature. Bharata began with the observation that the human being has a wide range of psychological states or emotions (*bhāva*s). He produced a list of forty-one such emotions, but did not give equal value to all forty-one. Eight among them have a dominant or durable (*sthāyin*) effect on the human personality. These are called the *sthāyi-bhāva*s. Dominance or durability, in this case, seems to be defined as those emotional states that are so engrossing, and affect the person feeling them so greatly, that for the time being that person forgets all else. The remainder of the forty-one emotions presumably lack this characteristic. The eight *sthāyi-bhāva*s listed by Bharata are passion (*rati*), humor (*hāsa*), anger (*krodha*), sorrow (śoka), effort (*utsaha*), astonishment (*vismaya*), disgust (*jugupsā*), and fear (*bhaya*).

Bharata maintained that these dominant emotions exist in the heart/mind of everyone in the form of unconscious latent impressions (*vāsanās*), which are derived either from actual experience in one's present life or from a previous existence. As such, they are ready to emerge into consciousness under the right conditions. Bharata then proceeded to determine by what means these various *sthāyi-bhāva*s could be evoked and raised to a conscious, relishable state.

His solution was seemingly simple: If an emotion arises in a certain environment and produces certain responses and gestures, cannot a representation of that environment, and an imitation of those responses and gestures, reproduce the emotion in the sensitive and cultured viewer (perhaps even in the actor)? Acting on this assumption, Bharata analyzed the emotions of ordinary life. This analysis revealed to him that emotions are manifested and accompanied by three components: the environmental conditions or causes (*kāraṇa*), the external responses or effects (*kārya*), and accompanying, supportive emotions (*sahakārin*). Bharata then went on to define the special characteristics of each of these components so that they could be imitated on stage and reproduce the desired emotion. However, when the environmental conditions, the external responses, and the accompanying emotional states are not part of real life, but are components of artistic expression divorced from the ordinary arena of personal egoistic concerns, they are technically renamed, respectively, the *vibhāva*s, the *anubhāva*s, and the *vyabhicāri-bhāva*s.

The *vibhāva* is generally explained as denoting that which makes the dominant emotion (*sthāyi-bhāva*) capable of being sensed. The later tradition recognizes a twofold division of the *vibhāva:* the substantial (*ālambana*), which consists of the actual characters of the play, and the enhancer (*uddīpana*), which is the setting of the play—garden, moon, and so forth.[9] The *anubhāva* is that which makes the *sthāyi-bhāva* actually sensed. The *anubhāva*s are the action of the play. They include words, gestures, and a group of involuntary physical responses (for example, trembling) called *sāttvika-bhāva*s.[10] The *vyabhicāri-bhāva*s are accessory emotions which foster, support, and give fresh impetus to the *sthāyi-bhāva.*[11]

By a carefully prescribed combination of these three main dramatic components, a *sthāyi-bhāva,* lying dormant as an unconscious impression, can be roused in the sympathetic spectator to a relishable state of aesthetic enjoyment—*rasa.* This is the meaning of Bharata's famous *rasa-sūtra:*

> *Rasa* is produced from the combination of the *vibhāva,* the *anubhāva,* and the *vyabhicāri-bhāva.*[12]

The eight *sthāyi-bhāva*s, when combined with the *vibhāva*s, *anubhāva*s and *vyabhicāri-bhāva*s in the controlled environment of the theatre, are experienced as the eight distinctive dramatic tastes or *rasa*s: amorous (*śṛṅgāra*); humorous (*hāsya*); furious (*raudra*); pathetic or compassionate (*karuṇa*); heroic (*vīra*); amazing (*adbhuta*); odious (*bībhatsa*); and terrible (*bhayānaka*). All later writers on *rasa* theory, concerned with either drama or poetry (these were theoretically treated as the same in India), center their discussions on the meaning of Bharata's *rasa-sūtra.* In the following chapter we will see the vital role it played in the religio-aesthetic system of Rūpa Gosvāmin.

The *sthāyi-bhāva*s, or dominant emotions, assume a very intense quality when experienced in the context of the theatre; however, this quality is quite different from emotions experienced in the personal context of individual egoistic concerns. Aesthetic experience, *rasa,* somehow lifts us out of the sorrowful conditions of everyday alienated experiences, which the *Nāṭya-śāstra* tells us are caused by egoistic emotional problems. The aesthetic experience is marked by a sense of wonder. M. Christopher Byrski argues that the sentiment of amazement, the *adbhuta rasa,* is to Bharata the culmination of all dramas, which occurs as one is lifted out of the world of personal emotional turmoil and experiences a sense of reintegration with the cosmic whole, the Absolute.[13] The religious implications of these ideas are clear, yet it was left to later writers to define explicitly the special "religious" quality of drama, which is at best only hinted at by Bharata. Let us therefore turn to some of the most prominent among these later writers.

Abhinavagupta's Comparison of Aesthetic and Religious Experience

The religious understanding of aesthetic experience in South Asia is primarily dependent upon a concept first mentioned by Bhaṭṭa Nāyaka,[14] a Kashmiri writer of the tenth century. Bhaṭṭa Nāyaka maintained that drama has a special power. "This power has the faculty of suppressing the thick layer of mental stupor (*moha*) occupying our own consciousness."[15] The effect of this power is the universalization or "generalization" (*sādhāraṇī-karaṇa*) of the emotional situation presented on stage. This theory of "generalization" requires further attention.

The *rasa* theorists insist that the experience of *rasa* is in all cases an enjoyable experience. Art is invariably delightful.[16] But how can that be in the instance of a drama that portrays, for example, grief? How is it that we can say that we "enjoyed" the tragedy of *Romeo and Juliet?* Bhaṭṭa Nāyaka contends that *rasa* is an experience somehow different from the direct personal experience of ordinary life. *Rasa* is generalized experience. Generalization is here understood to mean a process of idealization by which the sensitive viewer passes from his troublous personal emotion to the serene contemplation of a dramatic sentiment.[17] This process occurs through an identification with the impersonal situation. Generalization is thus a special state of identification with the world of dramatic representation, which transcends any practical interest or egoistic concerns of the limited self.

An ordinary emotion may be pleasurable or painful, but a dramatic sentiment (*rasa*), a shared emotion transcending personal attitude and concerns, is lifted above the pleasure and pain of personal ego into pure impersonal joy (*ānanda*). This happens because one is not concerned with how the depicted actions will personally affect one; an "artistic distance" is maintained between the spectator and the portrayed emotions. Aesthetic emotions are *intense,* but not personal. Juliet's painful situation does not affect our lives personally, but is ours to "enjoy" free from any of the ordinary concerns. Thus, the tears one sheds while watching a drama are never tears of pain, but of sentiment. Raniero Gnoli explains Bhaṭṭa Nāyaka's notion this way:

> During the aesthetic experience, the consciousness of the spectator is free from all practical desires. The spectacle is no longer felt in connexion with the empirical "I" of the spectator nor in connexion with any other particular individual; it is the power of abolishing the limited personality of the spectator, who regains, momentarily, his immaculate being not yet overshadowed by māyā.[18]

For the duration of the aesthetic experience, one steps out of ordinary time, space, and—most important of all—identity.

Bhaṭṭa Nāyaka was the first to develop an explicit explanation of aesthetic experience in terms of the spectator's inward experience. He suggests that the aesthetic experience of *rasa* is similar, though not identical, to the tasting (*āsvāda*) of the supreme *brahman.*[19] J. L. Masson and M. V. Patwardhan remark: "It may well be that Bhaṭṭa Nāyaka was the first person to make the famous comparison of yogic ecstacy and aesthetic experience."[20] Although the aesthetic experience for Bhaṭṭa Nāyaka is admittedly one of pure contemplation dissociated from all personal interests and results in composure (*viśrānti*), it is still marked by temporality and does not completely escape egoistic impulses since it is dependent upon the unconscious impressions (*vāsanā*s) which consist of acquired personal experience. These ideas of Bhaṭṭa Nāyaka had a tremendous influence on Abhinavagupta. Many of the themes which were later to occupy Abhinava are found in the remaining fragments of the works of Bhaṭṭa Nāyaka.

Before examining the aesthetic theory of Abhinavagupta, it will be useful to understand something of the religious context within which he was writing, for his comparison of aesthetic and religious experience assumes a particular notion of religious experience. Abhinava was deeply involved in the religious world of twelfth-century Kashmir Śaivism, which was informed by the philosophical system of Advaita Vedānta. He was one of the greatest philosophical minds of medieval Kashmir and considered an authority in all philosophical issues in Kashmir Śaivism. Two texts that were influential on his thought were the *Vījñānabhairava,* which is preoccupied with ecstatic experiences and exercises for inducing them, and the *Yogavāsiṣṭha–mahārāmāyaṇa,* a text which emphasizes (among many other things) unfettered enjoyment. The latter work urges all beings to strive for bliss (*ātmānanda*): "That is the highest place, the peaceful way (i.e. state), the eternal good, happiness (*śiva*). Delusion no longer disturbs the man who has found rest (*viśrānti*) there."[21] This foreshadows the terminology used by Abhinava to describe the aesthetic state.

The only description of Abhinavagupta that survives pictures him as a Tantric mystic.[22] Ritual plays a very important role in Kashmir Tantrism. The purpose of the Tantric ritual, according to the *Tantrāloka* of Abhinavagupta, is to "reveal" or "suggest" (*abhivyakti*) the blissful experience of the Self (*ātmānanda*).[23] Involvement with Tantric rituals affected Abhinava's views on the eventual goal of art, and led him to his transcendental theories of the aesthetic experience. The close connection in Abhinava's mind between the Tantric ritual and aesthetic experience is illustrated by the following quotation from Abhinava's *Tantrāloka:*

> The consciousness, which consists of, and is animated by, all things, on account of the difference of bodies, enters into a state of expansion—since all the components are reflected in each other. But, in public celebrations, it returns to a state of expansion—since all the components are reflected in each other. The radiance of one's own consciousness in ebullition (i.e., when it is tending to pour out of itself) is reflected in the consciousness of all bystanders, as if in so many mirrors, and, inflamed by these, it abandons without effort its state of individual contraction.

> For this very reason, in meetings of many people (at a performance of dancers, singers, etc.), fullness of joy occurs when every bystander, not only one of them, is identified with the spectacle. The consciousness, which, considered separately also, is innately made up of beatitude, attains, in these circumstances—during the execution of dances, etc.—a state of unity, and so enters into a state of beatitude which is full and perfect. In virtue of the absence of any cause for contraction, jealousy, envy, etc. the consciousness finds itself, in these circumstances, in a state of expansion, free of obstacles, and pervaded by beatitude. When, on the other hand, even one only of the bystanders does not concentrate on the spectacle he is looking at, and does not share, therefore, the form of consciousness in which the other spectators are immersed, this consciousness is disturbed, as at the touch of an uneven surface. This is the reason why, during the celebration of the *cakra,* etc., no individual must be allowed to enter who does not identify himself with the ceremonies and thus does not share the state of consciousness of the celebrants; this would cause, in fact, a contraction of the consciousness.[24]

According to De, Abhinavagupta's *rasa* theory is accepted as authoritative and adopted by all later writers on the subject. This assessment of Abhinava's theory is shared by many scholars of Indian aesthetics. "There can be little doubt," assert Masson and Patwardhan, "that Abhinava is the greatest name in Sanskrit literary criticism. For later writers on Sanskrit aesthetics there is no more important name than Abhinava."[25] These scholars would have us believe that all later *rasa* theorists agreed with Abhinava and accepted his position without question. This assertion, however, obscures the fact that many writers disagree with Abhinava on a number of major issues. After examining the historical debates regarding *rasa* theory, it would be difficult to maintain that Abhinavagupta's *rasa* theory is *the rasa* theory of India. Nevertheless, the claim could be made that Abhinavagupta's *rasa* theory was widely known and accepted. For this reason it will be useful to look at it closely, to judge how Rūpa Gosvāmin's understanding and use of the aesthetic principle of *rasa* differed from a commonly accepted understanding of the *rasa* experience.

Abhinava defines *rasa* as the very soul of drama or poetry:

> It belongs (*gocara*) only to the (suggestive) function in poetry. It is never included under worldly dealings (*vyavahāra*) and is never even to be dreamed of as being revealed directly through words. No, quite the contrary, it is *rasa,* that is, it has a form which is capable of being relished (*rasanīya*) through the function (*vyāpāra*) of personal aesthetic relish (*carvaṇā*), which is bliss (*ānanda*) that arises in the *sahṛdaya*'s delicate

> mind that has been colored (*anurāga*) by the appropriate (*samucita*) latent impressions (*vāsanā*) that are deeply embedded from long before (*prāk*); appropriate that is, to the beautiful *vibhāva*s and *anubhāva*s, and beautiful, again, because of their appeal to the heart (*saṃvāda*), and which are conveyed by means of words. That alone is *rasadhvani,* and that alone, in the strict sense of the word, is the soul of poetry.[26]

We observe in the above quotation that Abhinava took great interest in the *rasa* theory of Bharata and the *dhvani* theory of Ānandavardhana;[27] in fact, he wrote commentaries on both. He interpreted Bharata's *sūtra*—that *rasa* is produced from the combination of the *vibhāva, anubhāva,* and *vyabhicāri-bhāva*—to mean that *rasa* comes from the force of one's response to something that is already existing (as a lamp reveals an existing pot), not something that is produced. It is when the unconscious latent impressions are roused to consciousness in the theatre by the *vibhāva*s and so forth, and are responded to sympathetically, that one experiences *rasa.* The *nature* of one's response is particularly important for Abhinava. "Poetry," he tells us, "is like a woman in love and should be responded to with equal love."[28]

Abhinava maintained that one becomes receptive to a poem or drama by removing certain obstacles (*vighna*s). The aesthetic experience, for him, consists of a tasting (*āsvadā*) devoid of any of these obstacles; it is an undisturbed relish. Masson and Patwardhan comment: "All of Abhinava's efforts focus on one important need: to crack the hard shell of 'I' and allow to flow out the higher self which automatically identifies with everyone and everything around."[29] It seems, in fact, that all synonyms used for aesthetic pleasure are just other names for consciousness free of all obstacles. Moreover, for Abhinava the obstacles that hinder one from truly appreciating a poem or drama are the same obstacles that maintain the illusive "I," and thereby cause all ignorance and bondage in the Vedānta system of thought. The idea of Vedāntin liberation or *mokṣa,* which is manifest by the removal of enveloping obstacles, thus finds an analogy in the idea of the manifestation of *rasa.*

Abhinava's entire treatment of the *rasa* theory displays a deep concern with the parallels between aesthetic experience and the experience of the Vedāntin mystic. His justification for this comparison is quite evident and well illuminated in the following summary of his theory.

> Reduced to its bare essential the theory is as follows: watching a play or reading a poem for the sensitive reader (*sahṛdaya*) entails a loss of the sense of present time and space. All worldly considerations for the time being cease. Since we are not indifferent (*taṭastha*) to what is taking place,

> our involvement must be of a purer variety than we normally experience. We are not directly and personally involved, so the usual medley of desires and anxieties dissolve. Our hearts respond sympathetically (*hṛdāyasaṃvāda*) but not selfishly. Finally the response becomes total, all-engrossing, and we identify with the situation depicted (*tanmayībhavana*). The ego is transcended, and for the duration of the aesthetic experience, the normal waking "I" is suspended. Once this actually happens, we suddenly find that our responses are not like anything we have hitherto experienced, for now that all normal emotions are gone, now that the hard knot of "selfness" has been untied, we find ourselves in an unprecendented state of mental and emotional calm. The purity of our emotion and the intensity of it take us to a higher level of pleasure than we could know before—we experience sheer undifferentiated bliss (*ānandaikaghana*). . . . Inadvertently, says Abhinavagupta, we have arrived at the same inner terrain as that occupied by the mystic, though our aim was very different than his.[30]

Aesthetic experience, then, for Abhinava is similar to the mystic's experience (*brahmāsvāda*) in that both are uncommon (*alaukika*) experiences in which the self is forgotten. Abhinava reserves his greatest praise of the dramatic experience for that moment when the spectators so deeply enter into the world of the play that they transcend their own limited selves and arrive at the unity shared by the Vedāntin mystics. Moreover, both aesthetic and mystical experiences are brought about by the removal of obstacles. Present time and space disappear for the duration of the experience, and one is totally immersed in an experience marked by bliss (*ānanda*). Furthermore, Abhinava defines both the aesthetic and the religious experience with the Sanskrit term *camatkāra,* which means "wonder" or "astonishment" and implies "the cessation of a world—the ordinary, historical world, the *saṃsāra*—and its sudden replacement by a new dimension of reality."[31]

Abhinava maintained, however, that there are differences between the two kinds of experience. First, the aesthetic experience is characterized by temporality; the experience ends when one leaves the theatre. After the performance the members of the audience once again return to their separate selves. Drama is also not expected to change one's life radically. Abhinava could not say the same for the mystic's experience. The experience of *mokṣa* is much more profound, is very likely to make a drastic change in one's life, and necessarily becomes a permanent feature of life. Yet, more important, the two experiences are distinguished by the fact that, while the experience of *mokṣa* is by definition beyond illusion, the aesthetic experience still partakes in illusion.

Drama functions for Abhinava much like the dream state did for Śaṅkara. In the Vedānta system the dream state of consciousness (*svapna-sthāna*) is the second of the four levels of consciousness—that is, waking, dream, dreamless, and transcendental consciousness (*samādhi*). Śaṅkara observed that the stuff of dreams, although rooted in the waking state, is not empirically real in the same sense as the content of the waking life, for one recognizes it for what it is, namely, illusion. Eliot Deutsch writes: "No matter how deeply involved one is with the objects of dream, one retains an independence from them and indeed a greater freedom with respect to them than is possible in waking consciousness."[32] To Abhinava, this is also true in the aesthetic experience of drama. One is deeply involved in the objects on stage but is free from a direct entanglement with them because of the artistic distance. One does not question the reality of the objects; they are recognized as illusion. The aesthetic experience demonstrates concretely how one can enjoy illusory objects while not being bound by them. The result is the experience of bliss. Therefore, in the aesthetic experience, which, like philosophy, instructs, one moves one step closer to the state of freedom in which the illusive nature of even the objects of the waking state is perceived.[33] The goal of aesthetic "instruction" is the ability to sit back and truly enjoy the cosmic drama, *saṃsāra,* created by the ultimate playwright, Śiva: "to attain aesthetic bliss by watching the spectacle of the play that is our own life in this world."[34]

While Abhinava contends that during the aesthetic experience of a drama one is more free from illusion than in the waking state, we must remember that he places serious limitations on this experience. In the aesthetic experience one is still experiencing the binding emotional contents of the individual unconscious, the *vāsanā*s. Nevertheless, he believes that the aesthetic experience of drama can function as a pointer to that reality beyond illusion. "Art experience," remarks Mysore Hiriyanna, commenting on this issue, "is well adapted to arouse our interest in the ideal state by giving us a foretaste of it, and thus serves as a powerful incentive to the pursuit of that state."[35] It is on these terms that Abhinava says the purpose of drama is bliss or pleasure. "*Rasa* consists of pleasure, and *rasa* alone is drama, and drama alone is the Veda."[36] Abhinava means something very particular by bliss (*ānanda*); it is a bliss that instructs by placing us in a state of mental repose, in which the oppressive egoistic illusions can be shattered. Dramatic experience for Abhinavagupta is therefore a kind of cloudy window into a reality beyond illusion.

Major Issues

I have already mentioned that although Abhinava's theories seem to have met with general acceptance, by no means did they remain unchallenged. Differences on a number of major issues can be seen throughout the developmental history of *rasa* theories. To understand better the placement of Rūpa Gosvāmin's *rasa* theory in the history of this development, it will be necessary to examine briefly some of these issues.

Location of *Rasa*

One important issue which concerned the *rasa* theorists, particularly those who perceived the experience of *rasa* to be closely associated with religious experience, is the question: Who can experience *rasa?* That is, where is *rasa* located? This issue is never directly addressed by Bharata in the *Nāṭya-śāstra,* though he strongly suggests in my reading of him that only the cultured spectator (*sumanasaḥ preksakāḥ*)[37] "tastes" the dramatic *rasa.* Regardless, his lack of a definitive statement firmly identifying the location of *rasa* left much room for a variety of interpretations.

Bhaṭṭa Lollaṭa (ninth century) was the first commentator on Bharata's *rasa-sūtra* to address directly the issue of the location of *rasa.*[38] Interestingly, he states that "*rasa* is located in both the original character (*anukārya*) and also in the actor (*anukartarī*), due to the power of congruous connection (*anusandhāna*)."[39] Many contemporary scholars make much of Bhaṭṭa Lollaṭa's term *anusandhāna* and want to interpret it as the ability of an actor to identify with the role. Y. S. Walimbe, for example, writes:

> The emotion is also produced in the actor because of the strength of his identification with the original character. Thus, indirectly Bhaṭṭa Lollaṭa also underlines the necessity of the actor's identification with the role, without which his emotional experience is impossible.[40]

Gnoli explains *anusandhāna* as "the power thanks to which the actor 'becomes' for the time being the represented or imitated personage."[41] (If this is indeed the case, then Bhaṭṭa Lollaṭa's theory has interesting parallels with Constantin Stanislavski's theory of "reincarnation.") Bhaṭṭa Lollaṭa's theory thus represents one interpretive option. It concentrates on the experience of the actor and does not appear at all concerned with the experience of the spectator. Abhinavagupta strongly opposed this theory and, many will argue, buried it for all time (though I will demonstrate this to be a false contention).

Abhinavagupta refuses to grant the aesthetic experience of *rasa* to the actor.[42] The actor is too close, too technically involved, for Abhinava to permit him to have the experience of *rasa*. Instead, it is the spectator who is free to identify with the depicted situation and thereby experience *rasa*. "The fullness of the enjoyment depends essentially on the nature and experience of the spectator, to whom it falls to identify himself with the hero or other character, and thus to experience in ideal form his emotions and feelings."[43] Abhinava's own term for this identification with the situation depicted on stage is *tan-mayī-bhavana*. Aesthetic experience is dependent upon this identification. It is by means of this identification that the spectator leaves the time, space, and personal identity of saṃsāric existence and enters the generalized time, space, and identity defined by the drama. "The spectator is so wrapt in what he sees, so carried away by a mysterious delight (*camatkāra*), that he identifies completely with the original character and sees the whole world as he saw it."[44] Abhinava is insistent, however, that only the spectator, who has the proper "artistic distance" and can truly let go of the ordinary world, can experience the mysterious delight of *rasa*. Though he never states so directly, his theory that *rasa* exists only in art implies that the original character, along with the actor, is also denied the experience of *rasa*.

Two later writers of significant importance differ from Abhinava on this very issue. The first is Bhoja, an eleventh-century king of central India who wrote the *Śṛṅgāra Prakāśa*.[45] According to Bhoja, any cultured individual (*rasika*) can experience *rasa*. "The Rasika may be the spectator and the connoisseur, the poet, or the characters like Rāma in the story. . . . The actor who acts the character of the story is also Rasavān [i.e., possessing *rasa*]."[46] That is, one's position with respect to the drama does not necessarily determine whether one is capable of experiencing *rasa* or not. Rather, the condition of one's inner nature is the deciding factor. The ability to experience *rasa* depends upon the full bloom of one's emotional nature. A mature emotional condition produces the power to get into others' moods, the power of empathy. Bhoja clearly maintains that not all people are *rasika*s; one must come to such a condition by birth. The essential ingredient of a *rasika* is the quality of one's unconscious latent impressions (*vāsanā*s).[47]

The second writer who differs somewhat from the strict position established by Abhinava is Viśvanātha Kavirāja, a fourteenth-century writer from Eastern India who wrote the *Sāhitya-darpaṇa*.[48] Viśvanātha agrees with Abhinava that it is primarily the spectator who experiences *rasa*. He further agrees with Abhinava that the original character does not experience *rasa* because his or her emotion is worldly (*laukika*), whereas *rasa* transcends personal emotions (*alaukika*).[49] However, his views on the actor's experience appear to be a bridge between Abhinava's opinion (which refuses aesthetic experience to the actor) and the opinion of those who suggest that the actor's position is conducive to the experience of *rasa.* Viśvanātha agrees with Abhinava in stating: "Because of technical involvement in skills and practice, an actor who is representing the form of Rāma, etc., is not an experiencer of *rasa.*"[50] But the case is not closed there for Viśvanātha. "However," he goes on to write in the same verse, "by realizing the meaning of the drama, even he (the actor) is a spectator." The commentary clarifies this point: "If he realizes the meaning of the drama and identifies himself with his role, Rāma, etc. (*Rāmādi-svarūpatām ātmanaḥ*), then even the actor may be considered a spectator." The idea expressed here seems to be that the accomplished actor, who can transcend the mechanical nature of acting and move into the world of the play's perspective, can also experience *rasa.* Thus we witness not one but a variety of views regarding the location of *rasa* in the centuries between Bhaṭṭa Lollaṭa and Viśvanātha Kavirāja.

Relationship of the *Sthāyi-bhāva* and *Rasa*

The next important issue to be considered is the relationship between the *sthāyi-bhāva* and *rasa.* Bharata does not mention the *sthāyi-bhāva* in the *rasa-sūtra.* This lacuna has generated a significant amount of discussion. Elsewhere in the *Nāṭya-śāstra,* Bharata seems to assert that it is the *sthāyi-bhāva* that becomes *rasa* (*sthāyyeva tu raso bhavet,* 7.29). Still, what exactly Bharata means by this "becoming" is the subject of long debate.

Bhaṭṭa Lollaṭa wrote: "*Rasa* is simply a *sthāyi-bhāva* intensified by the *vibhāva, anubhāva,* etc.; but if it were not intensified it would remain a *sthāyi-bhāva.*"[51] Bhaṭṭa Lollaṭa thus maintains that there is a direct relationship between the *sthāyi-bhāva* and *rasa.* The difference between them is merely quantitative, not qualitative. One, the *rasa,* is only an intensified form of the other, the *sthāyi-bhāva.* The intensification occurs through contact with the *vibhāva, anubhāva,* and *vyabhicāri-bhāva.*

Bhaṭṭa Lollaṭa's views were first criticized by another ninth-century commentator on the *Nāṭya-śāstra,* Śaṅkuka.[52] Śaṅkuka argued that Bhaṭṭa Lollaṭa's theory of intensification assumed degrees of *rasa,* and that this violates the assertion expressed by Bharata that *rasa* is a homogeneous "taste." Śaṅkuka's own position is that *rasa* is not a *sthāyi-bhāva* at all; instead, it is an *imitation* of a *sthāyi-bhāva,* thus the new designation *rasa.* Śaṅkuka's theory of imitation, however, was not accepted by later writers.

Abhinavagupta, following Bhaṭṭa Nāyaka, insisted that *rasa* is a "non-worldly" (*alaukika*) experience that transcends ordinary emotions and exists only in art. He argued that the *sthāyi-bhāva,* on the other hand, is an emotion that exists in the everyday world. Therefore, he contended, *rasa* is quite different from a *sthāyi-bhāva.*[53] The *sthāyi-bhāva* is the experience of unconscious impressions, or *vāsanā*s, roused to consciousness in the everyday world of personal concern; *rasa* is the experience of the *vasānā*s roused to consciousness in the controlled and impersonal environment of the theatre. The *sthāyi-bhāva* belongs to the world, while *rasa* belongs to art; and for Abhinava, never the twain shall meet. It is for this reason, Abhinava argues, that Bharata did not mention the *sthāyi-bhāva* in the *rasa-sūtra.*

Once again, contrary to the assumption that Abhinava's position is *the* Indian position, we find later writers disagreeing with Abhinava on a very important issue. Both Bhoja and Viśvanātha Kavirāja create a theory of aesthetics based on a more direct understanding of the relationship that exists between the *sthāyi-bhāva* and *rasa. Rasa,* for Bhoja, is merely a manifest *sthāyi-bhāva.* He interprets Bharata's *sūtra* to mean that when the *vibhāva*s and other aesthetic components combine with and act upon the *sthāyi-bhāva, rasa* is produced; a developmental relationship is understood to exist between the *sthāyi-bhāva* and *rasa.* The similes Bhoja uses to explain the "production" (*niṣpatti*) of *rasa* from the *sthāyi-bhāva*s are of the production of juice from sugarcane, oil from sesame, butter from curds, and fire from wood.[54] Thus, Bhoja regards the *sthāyi-bhāva* and *rasa* "as fundamentally the same, different only in their designations (*jāti*), discharging different functions in reality (*arthakriyā*) and actually as so many stages (*avasthā*) of evolution of the same pattern."[55] In the initial stage there is *sthāyi-bhāva;* in the state of culmination there is *rasa.*

Viśvanātha also explains *rasa* as a development of the *sthāyi-bhāva.* He writes:

> The *sthāyi-bhāva* (passion, etc.) goes to the condition of *rasa* in the sensitive person when developed by the *vibhāva, anubhāva,* and *sañcārin* (*vyabhicāri-bhāva*).[56]

The commentary on this verse provides further explanation:

> *Rasa* is a manifestation developed (*pariṇata*) within the components like curds from milk. But it is not revealed as a previously existing pot is by a

lamp, as has been declared by the author of the *Locana* (i.e., by Abhinavagupta).

The above commentary points to one of the major controversies in the debate over the relationship between the *sthāyi-bhāva* and *rasa* among Indian aestheticians, and for that matter, between the nature of the "world" and Ultimate Reality among Indian religious philosophers. One group, the Pariṇāma-vādins, represented here by Bhoja and Viśvanātha, maintain that the world is a transformation or development (*pariṇāma*) of Ultimate Reality (*brahman*); whereas the second group, the Vivarta-vādins, represented here by Abhinavagupta, hold that the world is a false appearance (*vivarta*) of Ultimate Reality.[57] The Pariṇāma-vādins use the simile of the production of curds from milk to explain the existence of the world, whereas the Vivarta-vadins favor the analogy of a rope being mistaken for a snake, to explain the world's (false) existence. The perspectives of these two schools on the evaluation of aesthetic experience are clear. Pariṇāma-vādins see a developmental relationship existing between art (*rasa*) and the world (*sthāyi-bhāva*); ordinary emotions are simply an underdeveloped form of *rasa*. The Vivarta-vādins, on the other hand, insist that there is no direct correspondence between art (*rasa*) and the world (*sthāyi-bhāva*); art totally transcends the emotional experience of everyday life.

Number of *Rasa*s

The last issue I will examine before moving on to the *rasa* theory of Rūpa Gosvāmin is the number of *rasa*s.[58] This is an especially important issue, for while many continued to maintain the concept of multi-*rasa*s, most writers interested in the religious quality of the dramatic experience tended to single out one *rasa* among the many as the supreme culmination of all *rasa*s. A given religious assessment of aesthetic experience is then somehow dependent upon the nature of the one *rasa* chosen as supreme. I will briefly examine two major figures deeply involved in the endeavor to single out one particular *rasa* as special: Abhinavagupta and Bhoja.

Bharata had produced a list of eight *rasa*s in the *Nāṭya-śāstra*. Abhinavagupta, following others before him, added a ninth *rasa* to this list: *śānta rasa,* the tranquil sentiment. Abhinava's words on *śānta rasa* are extremely difficult, but it is the opinion of most scholars that, for Abhinava, this *rasa* is qualitatively different from the eight standard *rasa*s. In fact, this is the argument Abhinava provides to explain why

Bharata did not mention the *śānta rasa* along with the standard eight. The special place of the tranquil *rasa* in the thought of Abhinava perhaps can best be seen in his discussion of its *sthāyi-bhāva*. In the *Abhinavabhāratī,* his commentary on the *Nāṭya-śāstra,* he states that the knowledge of the truth (*tattva-jñāna*), the knowledge of the *ātman* (*ātma-jñāna*), or simply the *ātman* itself, is the *sthāyi-bhāva* of the *śānta rasa*. In light of what has just been said concerning his view of the relationship between the *sthāyi-bhāva* and *rasa,* the following passage illuminates the unique characteristics of this *rasa*.

> Therefore, the *Ātman* alone possessed of such pure qualities as knowledge, bliss, etc., and devoid of the enjoyment of imagined sense-objects, is the *sthāyibhāva* of *śānta*. Its status as a *sthāyibhāva* should not be explained in the same terms as the status, as a *sthāyibhāva,* in the case of other *sthāyibhāva*s (i.e. there is a great difference between the *Ātman*'s status as a *sthāyibhāva* and the other *sthāyibhāva*s). For *rati,* etc., which arise and disappear due to the emergence and disappearance of their respective causes, are called *sthāyibhāva*s in so far as they attach themselves for some time to the canvas (wall) in the form of the *ātman* which is of an unchanging nature relative to them. But knowledge of the truth is the canvas behind all emotions, and so it is the most stable of all the *sthāyibhāva*s. It transforms all the states of mind such as love, etc., into transitory feelings, and its status as a *sthāyibhāva,* having been established by its very nature, need not be specifically mentioned. And therefore it is not proper to count (knowledge of the truth) separately (in addition to the eight *sthāyibhāva*s). Between a lame bull and a dehorned bull, *bullness* (which is the generic property present in both bulls) is not considered as a third thing.[59]

The gist of the argument is this: The *ātman* is truly fundamental (*sthāyin*) compared to the eight standard *sthāyi-bhāva*s; it is the permanent foundation upon which all other *sthāyi-bhāva*s are formed. Compared to the *ātman,* the standard eight *sthāyi-bhāva*s are unstable (*vyabhicārin*). Therefore, the *sthāyi-bhāva* of *śānta rasa,* the *ātman,* is unique in that it does not belong to the world of ordinary emotions. Abhinava goes on to argue that Bharata did not mention *śānta rasa* and its *sthāyi-bhāva* because they belong to a higher plane of religious tranquillity (*śānti* or *viśrānti*) into which all *rasa*s ultimately resolve.[60]

Another final reduction of all *rasa*s into one supreme *rasa* is found in Bhoja's *Śṛṅgāra Prakāśa*. Therein he writes that, at an initial level, any of the emotions listed by Bharata (forty-nine: eight *sthāyi-bhāva*s, eight *sāttvika-bhāva*s, and thirty-three *vyabhicāri-bhāva*s) can become a *rasa*.[61] He continues to say, however, that finally all *rasa*s are based on the truly central and permanent ego (*ahaṃkāra*), and it is only by means of their association with the ego that the other emotions are enjoyed as *rasa*. Hence, all *rasa*s ultimately resolve into this *ahaṃkāra-rasa*, which Bhoja calls Love (*śṛṅgāra* or *prema*). V. Raghavan explains Bhoja's position this way:

> When the one Ahaṃkāra-rasa gets scattered into forty-nine and more emotional manifestations and each of them has attained a climax, there is again a synthesis. As the climax is reached, all Bhāvas become Preman or a kind of love from where they pass into the Ahaṃkāra-rasa. That is, Rasa is one. The names Rati, etc. pertain to the lower state of Bhāvanā. It is much below the state of Rasa, i.e., in the state called Bhāvanā, that the one Rasa gets into diverse forms with many delimiting characteristics. Beyond the path of Bhāvanā of definite and named Bhāvas is the experience of the bliss of Rasa in our own souls lit by the spark of Ahaṃkāra, as part of our very souls.[62]

Bhoja maintains that the amorous sentiment (*śṛṅgāra rasa*), originating from the dominant love-instinct perpetually associated with the soul and awakened by manifestations of beauty, is the ultimate source of all *rasa*s; thus it is the only *rasa*. We see then that both Abhinavagupta and Bhoja finally reduce all *rasa*s into one single and supreme *rasa* which, they claim, compares to the religious experience of *brahmāsvāda*.

These then are some of the major theories and issues as they came down to Rūpa Gosvāmin in the sixteenth century. We are now prepared to examine Rūpa's application of *rasa* theory to the religious environment of *bhakti* and determine where he stood on these major issues and what relevance they had to his theory of religious practice.

3

The Aesthetics of *Bhakti*

All glory be to the Moon (Kṛṣṇa),
Whose form is the nectar of all *rasas*.

RŪPA GOSVĀMIN, *Bhaktirasāmṛtasindhu*

The Gauḍīya Vaiṣṇava movement is generally associated with the career of the Bengali saint Caitanya (b. 1486 C.E.). Although Caitanya obviously inspired many of those with whom he came in contact, he left no writings; instead, he assigned the task of systematizing the tenets of the young movement to a group of theologians whom he had sent to reside in the North Indian town of Vṛndāvana.[1] These theologians came to be known as the Six Gosvāmins of Vṛndāvana. The literary efforts of this extremely influential group established the primary foundations of the sect. S. K. De correctly remarks: "It was the inspiration and teachings of the six pious and scholarly Gosvāmins which came to determine finally the doctrinal trend of Bengal Vaiṣṇavism which, however modified and supplemented in later times, dominated throughout its subsequent history."[2]

Thus the Six Gosvāmins deserve attention since it was their writings that determined the shape the religious system of Gauḍīya Vaiṣṇavism was to take. Three among the six are particularly distinguished for the quality and magnitude of their writing and its subsequent influence: two brothers, Sanātana and Rūpa Gosvāmin, and their younger brother's son, Jīva Gosvāmin. Of these, Rūpa Gosvāmin was most influential in establishing the theoretical foundation of Gauḍīya Vaiṣṇava religious practice.

In his efforts to delineate the practice of *bhakti,* Rūpa utilized the *rasa* theory of Bharata's *Nāṭya–śāstra.* One way to better understand Rūpa's distinctive application of this *rasa* theory to the religious situation of Vaiṣṇava *bhakti* is to compare his theory to that of Abhinavagupta, discussed in the previous chapter. Since Abhinava's theory was widely accepted in medieval India, such an exercise will help us understand how

Rūpa altered a popular understanding of the *rasa* theory to make it congruent with *bhakti.* Moreover, since Abhinava's interpretation of Bharata's *rasa* theory tends to be the most widely known among Western scholars, it provides a good comparative background for understanding the uniqueness of Rūpa's theory. There are also points of similarity between the two theories; Abhinava may even have influenced Rūpa. In their investigation of the works of Abhinavagupta, Masson and Patwardhan remark:

> It seems to us that the whole of the Bengal Vaiṣṇava school of poetics (and not only poetics, but philosophy as well) was heavily influenced by the teachings of Abhinavagupta and the tradition he follows, though nobody writing on the Bengal school has noticed this fact or tried to follow its lead. It is true that the Gosvāmins do not quote Abhinava directly, but we think his influence is quite clear.[3]

Although these two scholars offer no evidence to support their contention, it does seem quite plausible that the Kashmir school did have some influence on the Vṛndāvana Gosvāmins. Both Abhinava and Rūpa were seriously concerned with a religious understanding of aesthetic experience. In addition, some of Abhinava's key terms (such as *camatkāra*—used to describe the "wonder" of the *rasa* experience) feature significantly in the writings of Rūpa Gosvāmin. Abhinava's influence should not, however, be overestimated; other influences were equally strong (e.g., Bhoja and Viśvanātha). Furthermore, in comparing the religio-aesthetic theories of Rūpa and Abhinava, there are fundamental differences which must be acccounted for. These will come to light as we examine Rūpa's application of the *rasa* theory to *bhakti.*

Rūpa Gosvāmin's Application: *Bhakti Rasa*

In the *Bhaktirasāmṛtasindhu*[4] Rūpa asserts *bhakti* to be the one and absolute *rasa.* Rūpa was not the first to discuss *bhakti* in the context of aesthetic theory, although he was by far the most important. Abhinavagupta had mentioned *bhakti* in his discussion of *śānta rasa.*[5] To Abhinava, however, *bhakti* is not a separate *rasa;* he includes it in the list of emotions conducive to *śānta rasa.* The real pioneer work in presenting *bhakti* as a distinctive *rasa* is the *Muktāphala* of Vopadeva.[6] In the eleventh chapter of the *Muktāphala,* Vopadeva establishes that there are nine types of devotees or *bhakta*s (note that a *bhakta* is a person who posses *bhakti*), each associated with one of the nine *rasa*s (Bharata's eight, plus *śānta*).

The humorous (*hāsya*) and amorous (*śṛṅgāra*) *rasa*s are given particular attention. Detailed analysis is not provided; instead, Vopadeva simply illustrates each with quotations from the *Bhāgavata Purāṇa*. A contemporary of Vopadeva, Hemādri, wrote a commentary on the *Muktāphala* entitled the *Kaivalyadīpika,* wherein he furthered the work of Vopadeva by applying the various components of Bharata's *rasa-sūtra* to Vaiṣṇava *bhakti.*[7] The treatment is all too brief, but seems to have had seminal importance.[8] The means of attaining Viṣṇu are declared to be the *sthāyi-bhāva*s of *bhakti-rasa,* and the standard list of *sthāyi-bhāva*s of the nine *rasa*s is accepted. Viṣṇu and his *bhakta*s are listed as the substantial causes (*ālambana–vibhāva*s) of *bhakti-rasa,* and things related to Viṣṇu, such as his deeds, are the enhancing causes (*uddīpana–vibhāva*s). The traditional *anubhāva*s and *vyabhicāri–bhāva*s complete the treatment. But it was Rūpa Gosvāmin who was to give the detailed and sophisticated expression of *bhakti* in terms of the *rasa* theory that has remained, since the early sixteenth century, one of the most popular ways of speaking of *bhakti* in northern India.[9]

The early Vṛndāvana Gosvāmins began their theological speculations with the Upaniṣadic assertion that Ultimate Reality or God (Bhagavān) is existence (*sat*), consciousness (*cit*), and bliss (*ānanda*). In his *Bhagavat Sandarbha,* Jīva Gosvāmin provides insight into the distinctive Gauḍīya Vaiṣṇava view of Ultimate Reality—here understood as Bhagavān Kṛṣṇa—by distinguishing three aspects or powers of the essential nature of Kṛṣṇa (*svarūpa-śakti*) which correspond respectively to *sat-cit-ānanda: sandhinī-śakti* is the power of existence, which upholds life in the universe; *saṃvit-śakti* is the power of consciousness, which makes knowledge possible; and *hlādinī-śakti* is Kṛṣṇa's power of infinite bliss, by which he both experiences bliss and causes bliss in others.[10] The Gauḍīya Vaiṣṇavas hold this third power or energy to be the highest and most important aspect of Kṛṣṇa. And in the *Bhaktirasāmṛtasindhu,* Rūpa asserts that "that emotion called love (*rati*) is the play of the great power (*mahā-śakti,* which is the *hlādinī-śakti,* the power of infinite bliss) and participates in the inconceivable essential nature of God (*acintya-svarūpa*)" (*Bhaktirasāmṛtasindhu* 2.5.74; hereafter cited as BRS). Thus, love itself is identified as an aspect of the essential nature of God, and we witness the repetition of the famous claim that "God is Love."

The desired aim of Gauḍīya Vaiṣṇavism is to participate (*bhakti*) in this aspect of God, defined as love or infinite bliss. Since emotion was seen as the highest approach to the Kṛṣṇa, who reveals himself in a cosmic drama, Rūpa recognized the usefulness of the existing *rasa* theory to explain the process of *bhakti.* In fact, Kṛṣṇa himself came to be defined as *rasa.* The *Taittirīya Upaniṣad* had already equated *rasa* with Brahman (2.7: *raso vai saḥ*).[11] Rūpa continued this identification, understanding *rasa* now more in terms of Bharata's dramatics, and addresses Kṛṣṇa as the essence of all *rasa*s (BRS 1.1.1). Rūpa's task then was to devise a practice which would generate this love for Kṛṣṇa (Kṛṣṇa-rati) in the hearts of *bhakta*s and raise it to the supreme condition of *rasa.* In so doing, Rūpa presented religion as drama.

Rūpa installed love (*rati*), more specifically love for Kṛṣṇa (Kṛṣṇa-rati), as the dominant emotion or *sthāyi-bhāva* of his *bhakti* aesthetics. Under the right conditions, he declared, this love could be experienced as paramount bliss (*bhakti-rasa,* also called Kṛṣṇa-prema) in the heart of the sensitive one—now no longer the literary connoisseur, but the *bhakta.* In his typically scholastic fashion, Rūpa carefully outlined the various types of *bhakti-rasa.* At one level, Rūpa seems to accept Bharata's number of eight *rasa*s (BRS 2.5.114). But at a deeper level, like Bhoja he singles out the amorous sentiment as being clearly the most important. In fact, we will see that the amorous sentiment—*śṛṅgāra*—is the basis of all *bhakti-rasa*s, all of which are founded upon some form of the *sthāyi-bhāva* of love for Kṛṣṇa. Since all *bhakti-rasa*s are developed out of the same *sthāyi-bhāva* (Kṛṣṇa-rati), then all *bhakti-rasa*s are truly one. However, *bhakti-rasa* is experienced differently according to the different types of *bhakta*s. Rūpa arranges the resulting subdivisions of *bhakti-rasa*s into primary and secondary (BRS 2.5.115-116). The primary *bhakti-rasa*s, which are based directly upon the *sthāyi-bhāva* of love for Kṛṣṇa, are articulated in five subdivisions: tranquillity (*śānta*), servitude (*prīta,* [elsewhere, *dāsya*]), friendship (*preyas,* [elsewhere, *sākhya*]), parental affection (*vātsalya*), and amorousness (*madhura*). These, it should be noted, will correspond to the five types of exemplary characters and optional roles to be discussed in Chapter 4. The remaining seven *rasa*s are only *bhakti-rasa*s to the degree that they too are based on the *sthāyi-bhāva* of love for Kṛṣṇa, though in their case the relationship is indirect (BRS 2.5.39-46). The *sthāyi-bhāva* for the humorous (*hāsya*) *bhakti-rasa,* for example, is humorous love (*hāsa-rati*). No exemplary roles are based on these secondary *bhakti-rasa*s, however; they are seen rather to be supports for the five primary subdivisions of *bhakti-rasa.* Therefore, although Rūpa declares the number of *rasa*s to be eight, all eight finally resolve into one *bhakti-rasa* called Kṛṣṇa-prema, or Love, which is held to be the *rasa* par excellence (*bhakti–rasa–rāt*), and which is said to make the liberation of the Vedāntins (*mokṣa*) seem like straw.

To experience *bhakti-rasa,* the *bhakta* moves onto the stage of the drama which transforms the world. In Rūpa's religious system, Kṛṣṇa becomes the *bhakta*'s dramatic partner; he is the hero (*nāyaka*) of the ultimate play. The individual *bhakta* relates to him personally by dramatically taking a part in that play. The whole world, or at least all of Vraja (which, from the correct spiritual perspective, amounts to the same thing), becomes a stage on which to act out one's part; thus religion becomes drama and acting becomes a way of salvation. Rūpa needed a dramatic theory to describe his religious system, and such a theory was readily available. Utilizing the components of Bharata's *rasa* theory, Rūpa was able to express his interpretation of *bhakti* with added sophistication.

Rūpa outlines the aesthetic components common to all types of *bhakti-rasa*s in the second of the four major sections of the *Bhakti-rasāmṛtasindhu.* The various forms of Kṛṣṇa and his intimate companions (*parikara*s, or simply *bhakta*s) are presented as the substantial causes (*ālambana-vibhāva*s) of *bhakti-rasa.* These are the characters of the cosmic play. More specifically, Kṛṣṇa is listed as the object (*viṣaya*) of the love, and the intimate companions who populate the drama are listed as the subjects or vessels (*āśraya*s) of the love. The moon, forests of Vraja, Kṛṣṇa's flute, and other details are presented as the enhancing causes (*uddīpana-vibhāva*s) of *bhakti-rasa.* These are the settings of the religious scene. Once the *sthāyi-bhāva,* the special love for Kṛṣṇa, is established in the heart of the *bhakta,* the introduction of these *vibhāva*s, along with the other aesthetic components, intensify it and cause it to be experienced as *bhakti-rasa* (BRS 2.5.80).

The role of the individual *bhakta,* the religious action if you will, is governed and spelled out in terms of the dramatic concept of *anubhāva*s, the physical expressions of an inner emotion. Rūpa employs the additional terms of the *Nāṭya-śāstra*—the involuntary manifestations (*sāttvika-bhāva*s) and accompanying emotions (*vyabhicāri-bhāva*s)—to further define the emotional approach of *bhakti.*

In the two final sections of the *Bhaktirasāmṛtasindhu,* Rūpa goes on to explain the twelve subdivisions of *bhakti-rasa*s, describing the particular nature of each and its concomitant aesthetic component. For example, the object (*viṣaya*) of the *śānta* (tranquil) *bhakti-rasa* is Kṛṣṇa in his four-armed form, who is described as Paramātma, the crown jewel of the delight of the Self known as the highest Brahman.[12] The vessels (*āśraya*s) of this *rasa* are various types of ascetics. The particular religious settings, expressed as the enhancing causes (*uddīpana-vibhāva*s), are such things as the Upaniṣads, auspicious mountains, places of meditation, the Ganges, and so forth. The external expressions (*anubhāvas*) of the tranquil *bhakti-rasa* include fixing the eyes on the tip of the nose, rejecting ordinary actions, remaining silent, and sitting in meditative postures. Those emotions which accompany and foster (*vyabhicāri*) the tranquil *bhakti-rasa* are such feelings as indifference to worldly objects, ascetic firmness, transcendental joy, and disgust with the mundane. In a likewise and sometimes overwhelmingly detailed manner, each type of *bhakti-rasa* is illustrated with numerous scriptural quotes to aid in the understanding of its particular flavor.

Rūpa's work stands as a fascinating presentation of a religious life conceptualized in terms of dramatic theory; he used the language of Bharata's *Nāṭya-śāstra* to define the life of *bhakti*. However, as I have already mentioned, he interpreted Bharata's *rasa* theory in a very unique way in order to fit it into the religious context of Vaiṣṇava *bhakti*. It would now be fruitful to compare Rūpa's theory with the better known theory of Abhinavagupta to elucidate the distinctive nature of Rūpa's theory.

Distinctive Implications of Rūpa's Theory

Abhinavagupta was primarily concerned with the aesthetic experience of an audience involved in watching any good drama. Observable in the writings of Abhinava is a fascination with a special power of aesthetic experience. For him this experience involves a sympathetic identification (*tan-mayī-bhavana*) with a portrayed situation that has the ability to draw people out of their own everyday world. It is in this sense that dramatic experience is transcendental and therefore valuable for Abhinava. But there is no special emphasis on the drama *itself* for Abhinava. Rather, the emphasis is on the generalized experience drama can produce; any drama will do if it is technically good enough to produce this transcendental experience in the audience. Abhinava considered the dramatic world depicted on stage as a recognized fiction; it is, in fact, because it is *recognized* as fiction that it can produce truly free emotional experiences.

When we come to the *rasa* theory of Rūpa Gosvāmin, however, we find ourselves in an entirely different context. For Rūpa, there is only one drama that can produce true *rasa*—the divine play of Kṛṣṇa. When the analysis shifts to a single drama, which is held to be Ultimate Reality itself, significant changes result. The emphasis for Rūpa is not on the ability of generic drama to lift one out of everyday experience; rather, he is deeply concerned with the means by which one may participate in the one Real Drama. For the Gauḍīya Vaiṣṇava, salvation comes to be defined as an eternal participation in this absolute drama.

Therefore, while for Abhinavagupta, for whom the drama*s* are still illusory, dramatic experience only points to Ultimate Reality or is at best a hazy window into that domain, for Rūpa Gosvāmin it is the very doorway into Ultimate Reality. Abhinava developed only a rough analogy between aesthetic experience and the liberating experience; Rūpa claimed they are equal. As there is but one drama, there can be only one *rasa,* and this *rasa,* Rūpa exclaims throughout his works, is an aspect of God himself. "So close is the fit," remarks Gerow, "that we may wonder whether aesthetics became theology, or theology aesthetic."[13]

As we move from Abhinava's context of many staged dramas to Rūpa's one ultimate drama, we observe a concomitant shift of concern, from the passive experience of the audience to the active experience of the actor. In Rūpa's aesthetics of *bhakti,* direct participation in the dramatic world is greatly valued. *Rasa* is no longer experienced from the passive and removed position of the audience; rather, it demands active realization. With Rūpa, *rasa* becomes an experience generated through a deep and active involvement in the drama. We have seen that Abhinava refused to grant the experience of *rasa* to the actor; according to him, the actor is too close, too involved to experience *rasa.* Contrarily, in Rūpa's system, the *bhakta* experiences *rasa* to the exact degree that he or she is deeply engaged in Kṛṣṇa's drama in Vraja. The Gauḍīya Vaiṣṇava seeks participation in the drama of ultimate value. "The devotee by his ardent meditations not only seeks to visualise and make the whole Vṛndāvana-līlā of Kṛṣṇa live before him, but he enters into it imaginatively, and *by playing the part of a beloved of Kṛṣṇa,* he experiences the passionate feelings (*bhakti-rasa*) which are so vividly pictured in the literature."[14] Active participation is here stressed. *Bhakti* is an active religious tradition.[15] The emphasis is on direct personal experience. *Rasa* for Rūpa, in contrast to Abhinava, is not mere distant contemplation; it demands direct realization within the individual. Thus the *bhakta* becomes actor, for as Stanislavski has shown in our own day, it is the actor who is in the most favorable position to enter into the dramatic world and to experience that world as his or her own. "When an actor is inspired he is in the same natural and spontaneous state that is ours in life, and he lives the experiences and emotions of the character he portrays."[16]

Both Bhoja[17] and Viśvanātha allowed the actor the experience of *rasa* under certain ideal conditions, but neither of them granted special preference to the actor's position. Not since Bhaṭṭa Lollaṭa in the ninth century had anyone made such a claim; this seems to be Rūpa's innovation. Reemphasis on the actor, as opposed to the audience, constitutes one of the main contributions of Rūpa's *rasa* theory. The consequences of this shift in the theory are tremendous. Now the actor is allowed the loss of ordinary time, space, and identity, and a deep participation in the time, space and identity of the character being portrayed.[18] Therefore, it is the *bhakta* as "actor" who is judged to be in the best position to enter and participate in the dramatic world of Vraja and experience *bhakti-rasa*.

Rūpa's innovations did not end there; they also included an extension of aesthetic experience into all life. Abhinava, who was concerned with an actual staged drama, maintained that the aesthetic experience of *rasa* ended when the curtain falls on stage; it could not become a permanent feature of life. This is one of the characteristics that distinguishes the aesthetic experience, for Abhinava, from the ultimate experience of *mokṣa*. In contrast, Rūpa is singly concerned with a drama on which the curtain never falls; the Vraja-līlā is eternal. Accordingly, he sought a way to maintain the experience of *rasa* outside the walls of any theatre. The whole world ideally becomes a stage, and all beings potential actors in the ultimate drama of the Vraja-līlā. Ananda Coomaraswamy once wrote: "The best and most God-like way of living is to 'play' the game."[19] The only game worth playing, the only game that leads to Ultimate Reality for Rūpa, is taking a part in the Vraja-līlā revealed in the *Bhāgavata Purāṇa*. Thus, acting becomes a way of salvation.

To account for some of the differences in the aesthetic theories and resulting views of dramatic experience between Rūpa and Abhinava, it is useful to consider the different status granted the emotions in each scholar's own religious context. Gerow observes that "in Abhinavagupta's view the ultimate reality—*brahmāsvāda*—is still essentially different from emotionality, and not reachable through it. The analogy is suggested by the growing parallelism between religion and literature, but 'religion' is conceived through the intellectualized remoteness of the Vedānta, and is not the immediate and compelling vision of divine love of *The Gīta-govinda*."[20] In the monistic teaching of Vedānta there is no difference between the individual and the Absolute, nor between individuals. Thus it is generalized experience that interests Abhinava. Emotions imply differentiation of individuals, but in the final analysis all individuals are one and the same with the Absolute; only our ignorance keeps us from perceiving them as such. Aesthetic experience is therefore limited for Abhinava, because it is still dependent upon the stored memory impressions of illusory individual experience—the *vāsanās*. In the Vedānta system, these latent impressions are the very source of all emotions. For Abhinava, then, aesthetic experience seeks nothing more than the realization of those persistent modes of experience that are the agency of saṃsāric transmission. Hence, emotions, even when freed from the specific environment in which they normally come to be, cannot be realized as unrelated to personal experience, as is the *ātman* itself, and cannot therefore lead one to the Real.

In the system of the Gosvāmins, which maintains a position of differentiation within non-differentiation (*acintyabhedābheda*), the individual is real and separate from, while yet maintaining a sameness with, the Absolute. Here, moreover, personal experience is greatly valued. Rūpa frequently devaluates the Vedānta goal of union or *mokṣa* (BRS 1.1.4, 14, 17, 32, 34), for how could one have a relationship with Kṛṣṇa if individuality were given up? (The Vaiṣṇava speaks of tasting sugar, not becoming sugar.) The goal is not to lose individual being, but rather to overcome the ignorance that keeps us from realizing who we truly are. The aim of *bhakti* is the transformation of identity, not the Vedāntin identification with the non-differentiated One. This is one of the major differences between *bhakti* and Vedāntic Hinduism. Rūpa claims that one is ultimately a character in the Vraja-līlā—a servant, a friend, an elder, or more important, a lover of Kṛṣṇa's—but never Kṛṣṇa himself. The experience of love requires an object and a subject. Thus absorption into the Absolute is eschewed and an eternal emotional relationship with Kṛṣṇa is pursued. For the Gosvāmins, the removal of the fetters of *māyā* is possible only by means of *bhakti,* and the ideal emotions aroused by dramatic experience are the very essence of salvation. The source of these ideal emotions, however, is quite different in Rūpa's theory than in Abhinava's.

The emotions basic to aesthetic experience, the *sthāyi-bhāva*s, are common to all people, according to Abhinavagupta, and exist in the form of latent impressions (*vāsanā*s) waiting to rise into consciousness—that is, they are dependent upon previous ordinary experience and are merely manifested by the dramatic experience. Rūpa dissents on this point by claiming that the *sthāyi-bhāva* he has singled out for religious attention, love for Kṛṣṇa (Kṛṣṇa-rati), is an extremely rare, special, and particular kind of religious emotion; it does not exist naturally in the hearts of all, nor is it dependent upon the ordinary *vāsanās*.[21] Since Rūpa maintained that the special *sthāyi-bhāva* of love for Kṛṣṇa is not innate in all, he pictured it as a prize to be won. He explains that it "is born in two ways: either from the practice of *sādhana,* or for the very fortunate, by the grace of Kṛṣṇa and his *bhakta*s" (BRS 1.3.6). It may be suggested by poetic images of natural love, but it differs greatly from instinctive love. Ordinary love, dependent on impermanent objects, ultimately resolves into pain, but the unusual (*alaukika*) love for the inexhaustible Kṛṣṇa brings pure and unending pleasure. The assertion made here is that the *sthāyi-bhāva* of love for Kṛṣṇa is unique and far superior to the relative *sthāyi-bhāva*s that Abhinava refers to and, when cultivated to a relishable form, does not diminish as one departs from any theatre (For when the whole world is a theatre, where else is there to go?), but rather becomes a permanent feature of one's life.

Rūpa presents the relationship between the *sthāyi-bhāva* and *rasa* in the typical developmental manner of the Pariṇāma-vādins (BRS 1.4.1 and 2.5.73, 132).[22] He explains that once the special *sthāyi-bhāva* of love for Kṛṣṇa is planted in the heart of the *bhakta,* it goes to the position of *bhakti-rasa* or Kṛṣṇa-prema with the slightest introduction of the *vibhāva*s (e.g., an image of Kṛṣṇa, or the right moon), *anubhāva*s, and *vyabhicāri-bhāva*s (BRS 2.5.79 and 97). Thus, once the *sthāyi-bhāva* is present all else follows. The unique feature of Rūpa's system, however, is his notion that the *sthāyi-bhāva* of love for Kṛṣṇa, Kṛṣṇa-rati, is not readily available. This peculiar understanding of the *sthāyi-bhāva* results in a serious problem: If the *sthāyi-bhāva,* upon which rests the possibility of experiencing *bhakti-rasa,* is not innate in the hearts of all, whence does it come? This is an extremely important point. The key to Rūpa's entire religious system depends on the generation of the *sthāyi-bhāva* of this special love for Kṛṣṇa (assuming that one is not among those rare ones who receive it by unmerited grace). And how is this to be accomplished? It is to be accomplished by means of a particular religious method—a *sādhana.*[23] Rūpa's greatest task then was to devise a *sādhana* that would enable the *bhakta* to enter and participate in the dramatic world of the Vraja-līlā, thereby generating the essential *sthāyi-bhāva* of love for Kṛṣṇa, the foundation of the ultimately meaningful experience of *bhakti-rasa,* Kṛṣṇa-prema. The solution he came up with is a way of salvation conceptually based on acting: the Rāgānugā Bhakti Sādhana.

4

The Gauḍīya Vaiṣṇava Script and Its Exemplary Roles

Be unto me, O Lord, as a father to a son, as a friend to a friend, or as a lover to the beloved.

BHAGAVAD-GĪTĀ

In this chapter we examine the paradigmatic individuals that define the ultimate world of Gauḍīya Vaiṣṇavism. We begin with a brief look at the sociohistorical background of the early leaders of the Gauḍīya Vaiṣṇava movement, to help understand the context within which the Rāgānugā Bhakti Sādhana arose and to highlight the usefulness of paradigmatic individuals for desired religious transformations.

The Historical Context of Early Gaudīya Vaiṣṇavism and the Need for Transcendent Models

The entire northern portion of the subcontinent of South Asia was under Muslim political domination from the beginning of the thirteenth century. Muhammad Ghurī, a Muslim ruler from Afghanistan, invaded northern India with the intention of establishing a kingdom, defeated the Rajputs, under Pṛthvīrāj, at the second battle of Tarain in 1192, and took possesion of the kingdom of Delhi. In 1202, his Turkish general, Muhammad Bakhtyar Khilji, defeated Lakṣmaṇasena of Bengal, and the greater part of northern India was formed into a new political entity—the Delhi Sultanate. Muslim rule continued in northern India until its control was usurped by the British.

The sixteenth century opened with Sikandar Lodī in control of the region surrounding Delhi. The site of Vṛndāvana, located eighty miles south of Delhi on the west bank of the Yamunā River, and the surrounding area, known as Vraja, were included in his domain. His rule was a continuation of the Lodī Empire, which lasted until Babūr, the first Mughal ruler, defeated Ibrāhīm Lodī at the battle of Panipath in 1526. Alā-ud-dīn Husain Shāh ruled Bengal from the court at Gauḍa until the second Mughal ruler, Humāyūn, took control of that city in 1538. Therefore, northern India in the first half of the sixteenth century experienced a shift in political power—from the relatively strong Lodī and Husain Shāhi dynasties to the even stronger Muslim rule of the Mughal Empire. It was clear to all concerned that the Muslims had definite political control of northern India during this period. Recognizing this, Caitanya is reported to have said to the local Muslim ruler of Navadvīpa: "You are the qazi, and you have power over *hindu-dharma.*"[1] Edward C. Dimock points out that this statement expresses the "recognition of the potential power of repression that lies in the hands of the Muslims."[2]

Although the Muslims frequently refrained from exercising their potential repressive power and often had close political relations with Hindus, the mere existence of Muslim dominance produced a serious problem for the Hindu populace, since previous forms of Hinduism had been dependent on political control. For example, the type of Vaiṣṇavism expressed in the pre-Muslim *Viṣṇudharmottara* reveals a religious structure dependent on an image and temple established only by the authority of a *cakravartin,* a king who had never been defeated in battle and was considered to be Viṣṇu's representative on earth. Common images of Viṣṇu's *avatāra*s at this time were, significantly, the powerful boar (Varāha) and the mighty man-lion (Narasiṃha). What then happens when the Muslims topple these essential elements of the Vaiṣṇava structure? Hinduism must have undergone a substantial transformation with the establishment of Muslim rule and the subsequent loss of a political center.

One important aspect of this transformation involves the rise to popularity of the playful, amorous god, Kṛṣṇa-Gopāla. During the early foundation of Muslim political control, we can observe a shift of focus in Vaiṣṇava mythological concerns from the more warrior-like and kingly aspects of Viṣṇu to those of the passionate god of the forest.[3] Much of the lengthy Purāṇic history of this deity was curiously ignored after the Muslims had usurped political control. In the mythological narrative of the *Bhāgavata Purāṇa* Kṛṣṇa was born for the eventual purpose of killing his wicked uncle Kaṃsa and assuming his rightful place as king on the throne of Mathurā and later Dvārakā. In later religious literature, however, Kṛṣṇa's life in these two cities, the seat of his kingly rule and courtly life, does not seem to be important; it is primarily used to explain his absence from the *gopīs* of Vṛndāvana. The myth does not change, but a different part of it comes to be lived. It is as though the loss of the political center caused a retreat from that center as religiously significant and a shift of religious significance to another sphere of meaning.

Other scholars have noted this transformation. W. G. Archer observes that somewhere around the end of the twelfth century

> the Krishna story completely alters. It is not that the facts as given in the *Bhagavata Purana* are disputed. It is rather that the emphasis and viewpoint are changed. Krishna the prince and his consort Rukmini are relegated to the background and Krishna the cowherd lover brought sharply to the fore. Krishna is no longer regarded as having been born solely to kill a tyrant and rid the world of demons. His chief function now is to vindicate passion as the symbol of final union with God.[4]

Archer accounts for this shift in terms of the Muslim presence, though in a rather unconvincing manner. He argues that "romance as an actual experience became more difficult of attainment" under the Muslims. "Yet," he continues, "the need for romance remained and we can see in the prevalence of love-poetry a substitute for wishes repressed in actual life. It is precisely this role which the story of Krishna the cowherd lover now came to perform."[5] I would rather argue that the mythological shift of emphasis from hero-king to cowherd lover is a retreat from the political realm, not a retreat from actual romance. One wonders whether the Muslims controlled the bedrooms of the Hindus as much as they did their thrones. It is the warriorly and kingly aspects of Kṛṣṇa, aspects which dominated Vaiṣṇava mythology before the introduction of Muslim political control, that come to be overlooked in later Vaiṣṇavism, which tends to focus on Kṛṣṇa the amorous lover. Kṛṣṇa as lover is most certainly a pre-Muslim motif, but he becomes increasingly popular after the political center was lost to the Muslims. The cowherd lover is surely not a figure associated with the political center.

David Kinsley also notes that later Vaiṣṇavism focuses almost exclusively on Kṛṣṇa's "superfluous life as a youth in Vṛndāvana." Kinsley interprets this fact by explaining that Vṛndāvana is a world on the periphery—it lies outside the conventions of society, which must be abandoned in order to achieve liberation from the oppressive chains of existence.

> The drama, in the first place, takes place outside the normal confines of society. It occurs in a humble cowherd village or in the forests of Vṛndāvana. Indeed, the whole affair is out of this world. It is only after

> Kṛṣṇa has left Vṛndāvana that he begins to assume social responsibilities and it is only in Vṛndāvana that he is adored in the subsequent Kṛṣṇa cults.[6]

Perhaps Kinsley is right, but such notions were not particularly important for the dominant form of Vaiṣṇavism before the establishment of Muslim rule. Kinsley fails to contextualize his study and therefore cannot account for the cause of this shift. In the context of Muslim presence, this shift in the mythological concerns of the Vaiṣṇavas emerges as a gradual retreat from the Muslim-dominated sociopolitical center as a sphere of religious meaning and as an expansion of a relatively unstressed pre-Muslim province of meaning.

Hindu scripture makes it clear that there are problems for any Hindu living in a social system that fails to reflect Hindu *dharma.* Yet that is exactly what many Hindus at the beginning of the sixteenth century were forced to do. If, however, there was little hope of regaining control of the political sphere, there was serious need for an expression of Hindu *dharma* that placed the world of significant meaning far beyond that sphere controlled by the Muslims. This is precisely what the early Gauḍīya Vaiṣṇava leaders provided. In his informative study, "Social Implications of the Gauḍīya Vaiṣṇava Movement," Joseph O'Connell contends that the Gauḍīya Vaiṣṇavas "systematically shift the notion of dharma for the age out of the realm of public order and into the realm of *prema-bhakti.*"[7] O'Connell convincingly argues that an impasse in medieval Bengal between the Muslims and the Hindus was solved by a devaluation of the sociopolitical world by the Hindus. To avoid further Hindu–Muslim conflict and insure a smoothly functioning state, the Muslim rulers encouraged the Hindu shift to a religious world of meaning that transcended the existing political order. This shift parallels the change of focus noted above in Vaiṣṇava mythology.

One has merely to look into the background of Sanātana, Rūpa, and Anupama, the father of Jīva, to understand that the early Gauḍīya Vaiṣṇava leaders were well aware of the sociopolitical situation of their time. All three were high ministers in the Muslim court at Gauḍa.[8] In the *Bhaktiratnākara,* Sanātana and Rūpa are said to have been very close ministers (*mahāmantrin*) to Husain Shāh and, before Caitanya renamed them, were known in the Muslim court as Sāker Malik and Dabir Khās, respectively.[9] A crisis of some sort arose between the Shāh and the two brothers and they left the court of Gauḍa. The historian R. C. Majumdar provides the following account of this incident from collected sources:

> When Husain asked Sanātana to accompany him during his Orissa expedition he flatly refused, saying, "You are going to desecrate Hindu temples and break images of Hindu gods; I cannot accompany you." The irate king threw him into prison but he managed to escape by bribing the guards. Either on account of this or for other reasons, both Rūpa and his brother Sanātana became apathetic towards worldly life and, on the advice of Śrī Chaitanya, both renounced the world and went to Vṛndāvana.[10]

Regardless of the level of accuracy in this account, evidence does suggest that Sanātana and Rūpa crossed a threshold beyond which they were no longer able to tolerate the tension between their ideal selves as formed by their Hindu heritage[11] and the roles they played in the immediate society. This tension was presumably equally intolerable for the Muslim Shāh, and their former social roles came to an end.

The early leaders of Gauḍīya Vaiṣṇavism typically express in their writings a deep dissatisfaction with their former, socially defined identities. In fact, O'Connell goes as far as to say: "They detested their former selves."[12] Along with this antipathy toward their former social selves, however, there is expression of optimism concerning the possibility of identity transformation. The early Gauḍīya Vaiṣṇavas express "a basic judgement about human character, that it is malleable. This judgement affirmed a principle of flexibility within a civilization in which human character and behavior ordinarily were ascribed by inherited occupation and social status, and by age and sex."[13] Such a judgment amounts to a declaration that people can and do break through the stereotypes and socially defined identities ascribed to them. This breakthrough is accomplished through religious action, which opens up an ultimate identity existing in a world of meaning that transcends the determinism of social identity and everyday experience.[14]

Upon leaving his position as minister, tradition maintains that Rūpa addressed the following Sanskrit verse to his brother Sanātana:

> Where, alas, is Ayodhyā, the kingdom of Rāma now? Its glories have disappeared. And where is the famous Mathurā of Kṛṣṇa? It also is devoid of its former splendor. Think of the fleeting nature of things and settle your course.[15]

This is a telling verse. Sizing up the political situation from his privileged position in the Muslim court at Gauḍa, Rūpa knew well the slim possibility of participating in these ideal mythological kingdoms at the sociopolitical level. Instead, he sought a new way to enter and participate in these meaningful worlds, and his encounter with the inspirational saint Caitanya supplied him with an example of a new way. Soon after meeting Caitanya, Rūpa was sent by him to Vṛndāvana to devise and establish a means, a *sādhana,* that would lead those interested away from an increasingly meaningless sociopolitical world and closer to the ideal mythological world expressed by the Purāṇas—a world which transcended that controlled by the Muslims. In effect, a process of "resocialization" was required of the early Gauḍīya Vaiṣṇavas. This is in keeping with the other schools of *bhakti,* which, in contrast to previous forms of Hinduism, as A. K. Ramanujan remarks, "do not believe that religion is something one is born with or into."[16] Thus a *method* was needed to open a way into the transcendent world. The Rāgānugā Bhakti Sādhana, first systematically presented by Rūpa Gosvāmin, was the answer.

From our introductory remarks, we would expect any method of entrance into a transcendental reality to include, necessarily, the prominent presence and use of paradigmatic individuals as models of and for that world. This is indeed the case. The path which leads out of the ordinary social reality, into the paramount reality reflected in the Vaiṣṇava Purāṇic mythology, makes extensive use of transcendent models. After briefly reviewing the Gauḍīya Vaiṣṇava conception of the transcendent world of ultimate meaning, we will examine these important figures and the manner in which they were analytically presented by the early Gauḍīya Vaiṣṇavas.

The Exemplary Script: Kṛṣṇa-līlā

Ultimate Reality for the Gauḍīya Vaiṣṇavas was revealed in the form of a cosmic drama. This drama is known as the Kṛṣṇa-līlā; its highest form is the Vraja-līlā.[17] The word "*līlā*" is usually translated as "play."[18] Both meanings of the English term are applicable; Kṛṣṇa's *līlā* is both a dramatic performance and an expression of his unpredictable playfulness. The purpose of this playful drama, this divine revelation, is to provide humans with a model of, and for, perfection. "Kṛṣṇa manifests himself in the earthly Vṛndāvana in order to show people the way of *rāga:* 'He manifested the *rāgamārga* . . . he taught it by his *līlā.*'"[19] This *līlā* goes on eternally, but the Gauḍīya Vaiṣṇavas maintain that it was revealed on earth in historical time. "It is important to note that the Vṛndāvana-līlā is not a mere symbol or divine allegory, but a literal fact of religious history."[20] This exemplary historical event is recorded in scripture, particularly in the *Bhāgavata Purāṇa,*[21] in the form of a narrative. The narrative relates the entire life of Ultimate Reality in its highest personal form, the cowherd Kṛṣṇa. Enough has been written on the Kṛṣṇa-līlā that the details need not be repeated here,[22] except for mention of a particular feature considered important from the perspective of the Gauḍīya Vaiṣṇava.

The dramatic narrative includes not only Kṛṣṇa, frequently described as an actor,[23] but also his intimate companions (*parikaras*). The Gauḍīya Vaiṣṇavas focus particular attention on the activities of the youthful cowherd god who grew up in the enchanted forests of Vraja, surrounded by affectionate servants, playfully herding cattle with his male companions, endearing himself to his elders with mischievious pranks, and meeting secretly under the midnight moon with the adolescent cowgirls (*gopīs*) of the village for bouts of passionate love. The importance of the narrative lies in the various emotional relationships with Kṛṣṇa that are exemplified. According to Gauḍīya Vaiṣṇava thought, these relationships are possible because of a special quality of Kṛṣṇa; he has the ability to conceal his divine nature. The inhabitants of Vraja are unaware of Kṛṣṇa's awesome and majestic form, his *aiśvarya-rūpa,* the form revealed to Arjuna in the *Bhagavad-gītā.*

In the eleventh chapter of the *Bhagavad-gītā* we read of the results of an encounter with such a form. Arjuna, who has previously related to Kṛṣṇa as a close companion, says to Kṛṣṇa: "I desire to see your majestic form (*aiśvarya-rūpa*)."[24] Kṛṣṇa grants Arjuna his wish by revealing this form to him. Arjuna's response is one of awe and terror, and the close affection he once felt for Kṛṣṇa leaves him. Drawing away in fright, Arjuna begs Kṛṣṇa to return to his human form and resume an affectionate relationship toward him. Kṛṣṇa complies, assuming his gentle human form (*saumyam mānuṣam rūpam*), and once again Arjuna can relate to him as an intimate. What makes emotional relationships with the Godhead possible is the concealment of the awesome form by the gentle human form. In the language of Rudolf Otto, the *mysterium fascinans* dominates the *mysterium tremendum.*[25] This, say the Gauḍīya Vaiṣṇavas, is the distinctive characteristic of the relationships between Kṛṣṇa and the highest of the exemplary models, the inhabitants of Vraja (Vrajaloka). Kṛṣṇa appears to them in a sweet, lovely and infinitely approachable human form called the *mādhurya-rūpa,* the form most conducive to attraction and love. This quality of Kṛṣṇa leads to a closeness devoid of any hesitation. Its functioning in Vraja is illustrated by the following incident involving Kṛṣṇa and his mother, Yaśodā.

In the eighth section of the tenth book of the *Bhāgavata Purāṇa,* we read how Yaśodā, suspecting that baby Kṛṣṇa had eaten mud, peers into his tiny mouth. There she sees the entire universe, and realizing that her infant is not the helpless creature she presumed him to be, she is instantly seized with terror. All feelings of affection are overcome and replaced by a distant awe. Kṛṣṇa, however, soon causes her to forget this awesome form and once again assumes his sweet form, with the result that:

> Instantly the *gopī* (Yaśodā) lost memory of this and placed her son on her lap. Her heart was once again filled with intense affection.[25]

The revelation of the majestic form (*aiśvarya-rūpa*) took away the possibility of an intimate emotional relationship, whereupon its concealment within the sweet human form (*mādhurya-rūpa*) enabled affection to return. The Gauḍīya Vaiṣṇavas developed the distinction of the two forms to a doctrinal level and proceeded to analyze the various exemplary figures presented in the *Bhāgavata Purāṇa* according to their awareness of these forms. Those who were aware only of the sweet human form of Kṛṣṇa tended to see him as their very own (*mamatā*) and were thus able to establish a closer relationship with him.

The religious goal of Gauḍīya Vaiṣṇavism—union with the Ultimate Reality of Kṛṣṇa through love—is conceived of as an eternal participation in the emotional world of the Vraja-līlā. Because the path to participation in that world is heavily dependent on the use of paradigmatic individuals, it is important to examine these exemplary models as they were presented by the Vṛndāvana Gosvāmins, Sanātana and Rūpa.

The Transcendent Models

The Paradigmatic Individuals in the *Bṛhad–Bhāgavatāmṛta*

The *Bṛhad-bhāgavatāmṛta,*[27] written by Sanātana Gosvāmin, is one of the first substantial works of the Vṛndāvana Gosvāmins. The text is of major importance and served as an inspiration for subsequent works. One scholar in modern-day Vṛndāvana told me that Sanātana's works contained the "raw experiential energy" behind the system of the Vṛndāvana Gosvāmins, which was then further developed on aesthetic grounds by Rūpa and on philosophic grounds by Jīva.[28] This text is therefore a good starting place for our investigation of that system. As the title (translated, "The Essence of the *Bhāgavata Purāṇa*") suggests, the text proposes to delve into the true inner meaning of the *Bhāgavata Purāṇa,* with the intention of exploring the ideal world therein presented. Significantly, Sanātana accomplishes this task in the text by presenting and analyzing, in a hierarchical fashion, the exemplary roles available to the aspirant. The text is organized into two related sections, each constituting a separate narrative. The first, which most clearly outlines the exemplary models of *bhakti,* proceeds as follows.

After listening to the great sage Jaimini narrate the entire *Mahābhārata,* King Janamejaya desires to learn more (!). He requests Jaimini to tell him about the highest spiritual joy (which, for this text, is the result of an amorous relationship with Kṛṣṇa called *madhura-rasa*). Jaimini responds with another story. He relates that after Parīkṣit had heard the entire *Bhāgavata Purāṇa* from Śukadeva, he attained liberation and was flooded with love for Kṛṣṇa. Just prior to his departure from this world, his mother, Uttarā, begged him to give her a taste of the essential nectar (*amṛta*) of the *Bhāgavata.* Parīkṣit happily complied with this tale.

Once upon a time many great sages convened at the Daśāśvamedha ghat. During this auspicious gathering a devoted *brahman* householder approached them and, having first worshipped Kṛṣṇa in the form of a *śalagrāma* stone, honored the sages by feeding them and washing their feet. He then offered all the fruits of this service to Kṛṣṇa. The celestial sage Nārada, who happened to be among the sages, was so moved by this pious *brahman* that he sprang to his feet and praised the actions of the *brahman,* declaring that he was surely the greatest *bhakta* (devotee) of Kṛṣṇa. Humbly denying such a position for himself, the *brahman* told Nārada of a certain king living in the south who was a better *bhakta* and a true recipient of Kṛṣṇa's grace (*kṛṣṇa-kṛpā-pātra*—"*pātra*" literally means "vessel," but is also used to designate the actor of a drama). No calamity exists in this king's land, claimed the *brahman,* and all his subjects are devoted to Kṛṣṇa.[29] Upon hearing this, Nārada ventures south in a journey that develops into a great quest for the highest *bhakta* of Kṛṣṇa. During the course of his adventures, he encounters a host of individuals who are paradigmatic for the tradition.

When Nārada meets the southern king, he lauds him as the best of the *bhakta*s, but the king declines this position. He suggests that Indra, king of the gods, occupies that station. This ends the first chapter, entitled "Praise of Those Associated with the Earth." In the following chapter, "Praise of the Heavenly," Nārada leaves this world in his search for the perfect *bhakta* and the models become "transcendent."

Nārada travels to Indra's world and eulogizes the king of the gods, who immediately disqualifies himself by confessing pride in his high and royal position. Indra proposes that Nārada seek out Brahmā, who must certainly be a greater favorite of Kṛṣṇa. Accordingly, Nārada proceeds to the world of Brahmā, where sages are performing Vedic sacrifices.

There Nārada confronts Brahmā and declares him the most favored *bhakta.* Hearing this, Brahmā closes his eight ear-holes with his fingers and swiftly denies the statement. He remarks that he possesses little *bhakti* and that the worship he performs to Kṛṣṇa is done for the purpose of attaining *mukti* or "liberation" (the goal of the Advaitins, which in this text is devalued as being secondary to *bhakti*). Brahmā suggests to Nārada that Śiva is really the model *bhakta* he is seeking. Nārada than expectantly leaves for Kailāsa, the snowy abode of the ascetic Śiva. But before Nārada has an opportunity to blurt out his usual praise, Śiva informs him that he is unworthy of being even the servant of a servant of Kṛṣṇa. He confesses that he has the power to attain only *mukti* devoid of *bhakti.* Śiva kindly reveals to Nārada that the highest *bhakta* is not to be found among the gods, but among Kṛṣṇa's "dear ones" (*kṛṣṇa-priya-jana*). He nominates Prahlāda as the most fortunate and supreme model of *bhaktas.* This is enough to send Nārada off in the direction of Prahlāda, the son of Hiraṇyakaśipu, whose devotion brought down Viṣṇu's man-lion incarnation.

The next chapter, "Praise of the Bhaktas," begins with Nārada approaching Prahlāda and declaring his journey to be at an end. Prahlāda, however, manages to prevent Nārada from heaping too much praise on him and tells Nārada that the one truly deserving such praise is Hanumān, who, it is well known, embodies an exemplary attitude of servitude. The narrative continues with Nārada's encounter with Hanumān, who predictably eschews the praise and transfers it to the heroes of the *Mahābhārata,* the Pāṇḍavas, since they love Kṛṣṇa as a friend.

The following chapter, "Praise of the Dear Ones," opens with Nārada addressing the Pāṇḍavas with extreme admiration. Surely they must be Kṛṣṇa's favorites and therefore his greatest *bhakta*s. But they too refuse such praise. They single out the Yādavas of Dvārakā as worthy, because of their intimate friendship with Kṛṣṇa. In the city of Dvārakā, the Yādavas accept some of Nārada's praise, but claim that their virtue is due primarily to Uddhava, who accompanies Kṛṣṇa during his travels, acting as a close friend and advisor. Nārada excitedly confronts Uddhava with this knowledge. Uddhava humbly accepts Nārada's praise, but insists that his status as a *bhakta* is nothing compared to that of the inhabitants of Vraja. At long last we arrive at the supreme land of Vraja and the action begins to move more rapidly.

"Praise of the Full Ones" is the title of the final chapter of the first section. This chapter features the staging of a drama of the Vraja-līlā to relieve the distress of Kṛṣṇa, who is at this time residing in Dvārakā far from his loved ones in Vraja. The drama is intended to demonstrate the loving affection of the inhabitants of Vraja. Nanda and Yaśodā are praised for the quality of their parental affection. The highest praise of all, however, is reserved for the adolescent cowgirls of Vraja, the *gopīs*, because of the intensity of their passionate love, which cannot bear separation from Kṛṣṇa even for a moment. Nārada triumphantly declares to Kṛṣṇa: "Now my efforts have been fruitful and I have obtained knowledge of the people who are vessels (*pātras*) of your mighty grace."[30] Parīkṣit, the narrator of this story, emphasizes the main point of the tale by advising his mother, Uttarā, to pursue the adolescent cowboy (Kṛṣṇa) with a love identical to that of the *gopīs*, who have been marked as the very highest of the exemplary *bhaktas*.[31] So ends the first half of the text.

Sanātana presents the models of *bhakti* in a hierarchical manner. He begins his narrative-like analysis with the examination and portrayal of the religious life of an ideal householder. He then provides a glimpse of the ideal ruler and his exemplary acts. Yet these roles are not greatly valued, for they are not located in the ideal transcendental world. Sanātana soon leaves this world in search of the most worthy models for the life of *bhakti*. The next world, the abode of the gods, is inhabited by those following the way of knowledge (*jñāna*) and Vedic sacrifices (*karma*). Indra is quickly passed over as the highest model owing to his royal status. Brahmā, associated with the way of knowledge and Vedic sacrifices, and Śiva, who is the model *yogin,* are denied the title of the paramount model for *bhakti* because they are too involved in the Advaitin goal of *mukti,* which, in the system of the Gosvāmins, is the last obstacle to be overcome in order for one to possess true *bhakti.* The *bhakta* wants to taste sugar, not become sugar.

It is only in the fourth chapter, "Praise of the Bhaktas," that we begin to find an acceptable model for the world of *bhakti.* Prahlāda, who surrendered all in total faith to Viṣṇu, is mentioned as a model of general *bhakti.* Slightly above him is Hanumān, who selflessly serves Rāma and, thereby, exemplifies what is referred to in the text as servitude (*dāsya-bhakti*). One stage higher than servitude is friendship (*sākhya-bhakti*). Here the Pāṇḍavas, Kṛṣṇa's companions in the great war, are mentioned, but more important are the Yādavas, the companions of Kṛṣṇa in Vraja who later accompanied him to Dvārakā. Even more superior than these are the foster parents of Kṛṣṇa, Nanda and Yaśodā, who occupy the emotional position of parental affection (*vātsalya-bhakti*). Finally, however, it is the young women who possess a deep amorous attachment (*mādhurya-bhakti*) to Kṛṣṇa who are deemed the highest models for *bhakti.* The queens of Dvārakā, who are married to

Kṛṣṇa, are among these, but the most exalted models are the adolescent cowgirls of Vraja. Married to others, they risk all to meet with their beloved Kṛṣṇa in the beautiful forests of Vṛndāvana so that they might please him in acts of total love. Such are the roles that represent the ideal religious world as presented by Sanātana Gosvāmin in the *Bṛhad-bhāgavatāmṛta.*

The second division of Sanātana's text parallels the first, but the focus is shifted from the model *bhakta*s to the forms in which Kṛṣṇa, as the Ultimate Reality, manifests himself to the various types of *bhakta*s. Nārada embarks on yet another quest, this time in search of the place and form in which Kṛṣṇa reveals most fully. Each model present in the first division is associated in the second with a concomitant perspective of Ultimate Reality. The two divisions are in agreement, for it is in Vraja, as the charming lover of the *gopī*s, that Kṛṣṇa is declared to be the most complete manifestation of Ultimate Reality.

The interconnected relationships between the exemplary models and Kṛṣṇa are developed more clearly, and in greater detail, in Rūpa Gosvāmin's *Bhaktirasāmṛtasindhu.* From this work comes the Gauḍīya Vaiṣṇava notion that Ultimate Reality, though one, is perceived differently by each type of aspirant. Jīva Gosvāmin explains this tenet most completely in his *Bhagavat Sandarbha.* Beginning with the Upaniṣadic assertion that Ultimate Reality is one, he argues further that that one reality is perceived according to the particular capacity and stage of realization of the aspirant (*upāsaka–yogyatā vaiśiṣṭyena*).[32] This is the Vaiṣṇava version of "different strokes for different folks." Every individual has a different set of karmic predicaments to work out, hence each must assume a different role in accomplishing the task of salvation. Ultimate Reality, therefore, is revealed to each in a unique manner.[33] For those inclined to the path of knowledge, Ultimate Reality is grasped as the undifferentiated Absolute Ground of Being (Brahman); for those engaged in *yoga,* the divine is experienced as the Inward Regulator (Paramātman as *antaryāmin*); while those attached to the path of *bhakti* (here viewed as supreme because it both encompasses and surpasses the previous two) encounter Ultimate Reality as the infinitely qualified Divine Person (Bhagavān Kṛṣṇa), who is perceived specifically according to the type of relationship one has with Kṛṣṇa.

The Paradigmatic Individuals in the *Bhaktirasāmṛtasindhu*

The paradigmatic models presented in narrative form in Sanātana's *Bṛhad-bhāgavatāmṛta* are analyzed in a much more systematic manner in Rūpa's *Bhaktirasāmṛtasindhu.*[34] Two Sanskrit terms, important to Rūpa's presentation of the paradigmatic individuals of the world of *bhakti,* are *viṣaya* and *āśraya.* The terms are usually translated as "object" and "vessel" respectively, but here can best be understood in the context of drama. Shakespeare's *Romeo and Juliet* is a familiar example. Romeo's first sight of Juliet creates an intense love within him. In this scene Juliet is the "object," the *viṣaya* of love, and Romeo the "vessel," the *āśraya* of love. The *bhakta*s, who are paradigmatic for the Gauḍīya Vaiṣṇava tradition, are presented in the *Bhaktirasāmṛtasindhu* as the "vessels" and Kṛṣṇa is the sole "object," though he appears differently according to the unique capacity of each vessel.

Rūpa commences his analysis and presentation of the exemplary models by briefly reviewing a few negative figures among the scriptural models. We read in the *Bhāgavata Purāṇa* that even demonic characters, such as Kṛṣṇa's wicked uncle Kaṃsa, attained salvation due to the fact that their intense hatred kept them constantly preoccupied with Kṛṣṇa. Emotional ties of all kinds bind one to Kṛṣṇa and thus insure salvation. Yet Rūpa declares these models of hatred and fear to be unsuitable. "Because they are contrary to the friendly and favorable nature (*ānukūlya*) of *bhakti,* hatred and fear are ruled out" (BRS 1.2.276). Thus Kaṃsa and the many demons who haunt the forests of Vṛndāvana and threaten Kṛṣṇa are rejected as worthy models. Their emotions are not aesthetically appropriate to the mode of *bhakti.*

Rūpa's discussion of the models or vessels (*āsraya*s) of *bhakti* is located within his systematic presentation of the five possible *bhāva*s—the general types of emotional states or relational roles which characterize the ideal world. He maintains that the level of realization one can attain depends upon the quality of the vessel.

> Love becomes distinctive according to the quality of its vessel, as the reflection of a sunray becomes distinctive according to the different things, such as crystal, in which it is reflected. (BRS 2.5.7)

Thus an awareness of the qualities and position of the vessel or model one chooses as one's paradigmatic individual becomes extremely important.

The first general role (*bhāva*) Rūpa presents is the pure state (*śuddha-bhāva,* also known as the peaceful state [*sānta-bhāva*]), which features a lack of emotional involvement with Kṛṣṇa. For this reason, it assumes the lowest position on the hierarchical scale. Rūpa places in this position the *yogin*s and followers of the path of knowledge (*jñāna-mārga*), previously discussed in Sanātana's *Bṛhad-bhāgavatāmṛta,* who seek a oneness with Ultimate Reality. Those located here are said to lack a "taste" for Kṛṣṇa

(BRS 2.5.21). The *bhakta*s insist that Ultimate Reality is not only consciousness but is also an object of consciousness. The goal of the *bhakta* is to relish Ultimate Reality aesthetically, not merely to become one with it. Rūpa does not exclude those who seek identification with Ultimate Reality from his system of salvation; he simply designates that they have made only the first step toward realization and devalues their goal in comparison with *bhakti.*[35] He asserts that their goal of union or liberation, *mukti,* is like a river compared to the great ocean of *bhakti* (BRS 1.1.4). These ascetics are seen by Rūpa as being concerned with the happiness of the "self," not with the happiness of and from Bhagavān (BRS 3.1.5). Moreover, he writes that these are "devoid of even a trace of my-ness (*mamatā*)," one of the main characteristics of the highest *mādhurya* awareness. For this reason they cannot understand the highest *līlā* of Kṛṣṇa and cannot attain the supreme world of Vraja (BRS 2.5.9). Kṛṣṇa, as the object of religious consciousness (*viṣaya,*) is perceived by those on this level as Paramātman, and assumes the appearance of the four-armed Viṣṇu, the highest Brahman, slayer of enemies, tranquil and omnipresent. That is, he is here manifest in his majestic form (*aiśvarya-rūpa*). The two figures of the Vaiṣṇava mythological world that Rūpa lists as exemplary of this role are Sanaka and Sanandana, two ascetics found in the *Bhāgavata Purāṇa.* Since this role is not of serious consideration for Rūpa, he spends little time discussing it.

In Rupa's hierarchy, the next three emotional roles—servitude, friendship, and elderly affection[36]—are all connected to an increasing degree of "my-ness." The awareness of the awesome majesty of Kṛṣṇa is progressively replaced with an awareness of his sweet approachable nature, so distinctly present in the inhabitants of Vraja.

Rupa compiles long lists of exemplary figures which have been "taken from scripture or else commonly known" (BRS 3.3.52). The starred characters in the lists that follow have been singled out by Rūpa as the best of their class. These lists of names may be overwhelming for one unfamiliar with the mythological world of the Vaiṣṇavas, but I present them here to demonstrate the Gauḍīya Vaiṣṇava understanding of the ideal religious world in terms of a sophisticated and detailed role analysis.

The models of servitude (*prīti* or *dāsya bhāva*) are divided into two groups: the servants and the younger relatives. The four types of servants:

1. The chiefs: Brahmā, Śiva, and Indra.
2. The three types of dependents:
   (a) Those who beg to be spared: Kāliya the serpent, Jarāsandha the father-in-law of Kaṃsa, and the Jailor.

   (b) Those who are motivated by knowledge: Śaunaka, etc.
   (c) Those who are devoted to service: Candradhvaja, Harihaya, Bahulāśva, Ikṣāku, Śrutadeva, and Puṇḍarīka.
3. The attendants: Uddhava,* Dāruka, Jaitra, Śrutadeva, Śatrujit, Nanda, Upananda, and Bhadra.
4. The obedient companions:
   (a) In Dvārakā: Sucandra, Maṇḍana, Stamba, and Sutamba.
   (b) In Vraja: Raktaka,* Patraka, Patrī, Madhukaṇṭha, Madhuvrata, Rasāla, Suvilāsa, Premakanda, Makaradaka, Ānanda, Candrahāsa, Payoda, Bakula, Rasada, and Śārada.

In Vraja, Kṛṣṇa appears to his servants as two-armed; elsewhere he is sometimes two-armed and sometimes four-armed.

The two types of younger relatives:

1. The younger brothers: Sāraṇa, Gada, and Subhadra.
2. The sons: Pradyumna,* Cārudeṣṇa, and Sāmba.[37]

For these younger relatives, Ultimate Reality is perceived as Kṛṣṇa the Great Elder, who is strong and wise, and both protects and caresses. The knowledge of his majestic form is predominant outside of Vraja; those in Vraja are for the most part unaware of the awesome form.

The third emotional role for exemplary characters is that of friendship (*sākhya-bhāva*). Kṛṣṇa appears to these characters as a contemporary. While all other roles involve some type of inequality, his friends are defined as those *equal* to him, in form, qualities, and dress. One of the most important qualities of this role is intimate confidence (*viśrambha*); that is, the friends have no hesitation in approaching Kṛṣṇa as a close companion. These friends are divided into two camps: those residing in Dvārakā and those residing in Vraja.

1. The friends in Dvārakā: Arjuna,* Bhīmasena, Draupadī, and Śrīdāma.
2. The four types of friends in Vraja:
   (a) Friends of the heart—these friends have a trace of the affection of the elder for the younger, because they are slightly older than Kṛṣṇa and protect him from the wicked: Subhadra, Maṇḍalībhadra,* Bhadravardhana, Gobhaṭa, Yakṣendrabhaṭa, Bhadrāṅga, Vīrabhadra, Vijaya, and Balabhadra*.
   (b) Ordinary friends—these have a trace of servitude, since they are slightly younger than Kṛṣṇa: Viśāla, Aujasvī, Devaprastha,* Varūthapa, Maranda, Kusumāpīḍa, Maṇībandha, and Karandhama.
   (c) Dear friends—these are equal in age to Kṛṣṇa and thus possess pure friendship: Śrīdāma,* Sudāma, Dāma, Vasudāma, Kiṅkinī, Stokakṛṣṇa, Aṃśu, Bhadrasena, Vilāsī, Puṇḍarīka, Viṭanka, and Kalaviṅka.

(d) Dear friends who aid in the love affairs—these are considered to be the best of friends because they are keepers of Kṛṣṇa's deepest secrets: Subala,* Arjuna, Gandharva, Vasanta, and Ujjvala*.

The fourth category of characters involved in the Kṛṣṇa-līlā is comprised of those who possess the affection of an elder for a younger (*vātsalya-bhāva*). Kṛṣṇa's majesty is even less present here. Ultimate Reality appears to these as the dear child Kṛṣṇa with a ball of butter in his hand. The elders are of two types: the teachers of Kṛṣṇa and the caretakers. The exemplary figures Rūpa lists for this category are the queen of Vraja (Yaśodā)*, the king of Vraja (Nanda)*, Devakī, Vasudeva, Kuntī, Sāndīpana, and the various married *gopī*s who feed Kṛṣṇa.

The class of models which is by far the most important for the Gauḍīya Vaiṣṇava tradition, overshadowing all others, is made up of those women who exemplify the amorous sentiment (*mādhurya-bhāva*) toward Kṛṣṇa. These women know only the sweet form (*mādhurya-rūpa*) of Kṛṣṇa. Rūpa includes only an introductory examination of the emotional role of these in the *Bhaktirasāmṛtasindhu.* A detailed account of this role and the paradigmatic individuals associated with it takes us into another text written by Rūpa, the *Ujjvalanīlamaṇi,*[38] which is a sequel to the *Bhaktirasāmṛtasindhu.* Kṛṣṇa (as *viṣaya*) appears to these as the skillful lover and dramatic hero (*nāyaka*). The vessels (*āśraya*s) are the beloved women of Kṛṣṇa (*hari-priyā* or *hari-vallabhā*). They are divided into two groups: those who are married to Kṛṣṇa (*svakīyā*) and those who belong to another (*parakīyā*) (*Ujjvalanīlamaṇi* 3.3; hereafter cited as UNM). A third type of amorous relationship is mentioned by Rūpa later in the text: generalized love (*sādhāranī,* may also be understood as the love of a courtesan), which is exemplified by the hunchback Kubjā, who sought primarily to slake her own desires with Kṛṣṇa after he had straightened her back (UNM 14.43). The *svakīyā* relationship is marked with mutual satisfaction proper to marriage (*samañjasā*) and is portrayed by the queens of Dvārakā, particularly Rukmiṇī (UNM 14.43). Kṛṣṇa appears to these as a loving husband (UNM 1.11).

It is the *parakīyā* lover, however, that exemplifies the highest form of all relationships. For these, Kṛṣṇa is the most charming and sweetest of lovers (UNM 1.17). The *parakīyā* lovers are the *gopī*s, the adolescent cowgirls of Vraja whose erotic love is solely for the pleasure of Kṛṣṇa (*samarthā*). The *parakīyā* lover is of two kinds: an unmarried girl, and a married woman (UNM 3.19). It is the latter of the two, a woman married to another who risks all in her love for Kṛṣṇa, who is declared to be the deepest vessel, and the highest model of *bhakti* among all the exemplary figures previously listed. Rūpa names two types of *gopīs* who have eternally occupied this supreme position: those known from scripture (*śāstra-prasiddhā,* UNM 3.56–3.57), and those known from common knowledge (*loka-prasiddhā;* UNM 3.58).

1. Those known from scripture: Rādhā,* Candrāvalī,* Viśākhā, Lalitā, Śyāmā, Padmā, Śaivyā, Bhadrikā, Tārā, Vicitrā, Gopalī, Dhaniṣṭhā, and Pālikā.
2. Those known from common knowledge: Khañjajākṣī, Manoramā, Maṅgalā, Bimalā, Līlā, Kṛṣṇā, Śārī, Visāradā, Tārāvalī, Cakorākṣī, Saṅkarī, and Kuṅkumā.

Among these, Rādhā and Candrāvalī are said to be the best, as they are the leaders of groups of women (*yutheśvarī*); but finally Rādhā is declared to be the greatest of all (UNM 4.3). As a "vessel" of love, her depth is immeasurable.

The different types of female lovers (*nāyikās,* or dramatic heroines) are illustrated in the *Ujjvalanīlamaṇi* by the different stages and moods of Rādhā, but further detail is unnecessary for our purposes. One additional exemplary figure, however, must be mentioned, for we will see in later chapters that this figure was to assume a position of great importance in Gauḍīya Vaiṣṇava religious practice. Among the long list of Rādhā's qualities outlined in the *Ujjvalanīlamaṇi,* Rūpa includes the fact that she is a leader of friends (*sakhīs*). These important female friends, the *sakhīs*, are declared to be of five kinds:[39]

1. The ordinary friends (*sakhī*)—these have a greater love for Kṛṣṇa than for Rādhā: Kusumikā, Vindhyā, and Dhaniṣṭhā.
2. The eternal friends (*nitya-sakhī*)—these have a greater love for Rādhā than for Kṛṣṇa: Kastūrī and Maṇimañjarī.
3. The friends close as life (*prāṇa-sakhī*)—these too have a greater love for Rādhā than for Kṛṣṇa: Śaśimukhī, Vāsantī, and Lāsikā.
4. The dear friends (*priya-sakhī*)—these have an equal love for Rādhā and Kṛṣṇa: Kuraṅgākṣī, Sumadhyā, Madanālasā, Kamalā, Mādhurī, Mañjukeśī, Kandarpasundarī, Mādharī, Mālatī, Kāmalatā, and Śaśikalā.
5. The best friends (*parama-preṣṭha-sakhī*)—these too have an equal love for Rādhā and Kṛṣṇa. They are the eight traditional *sakhīs* (*aṣṭa-sakhīs*): Lalitā, Savisākhikā (Viśākhā), Sucitrā, Campakalatā, Tuṅgavidhyā, Indulekhikā, Raṅgadevī, and Sudevī.

These, then, are the transcendent models Rūpa presents as representative of the ideal religious world. Through these models, the entire emotional world of the Kṛṣṇa-līlā is portrayed and made accessible. In the following chapters we will examine in detail what use is made of these transcendent models or paradigmatic individuals in the quest to enter the world they inhabit.

Worthiness of the Paradigmatic Individuals

In my introductory remarks I stated that exemplary models or paradigmatic individuals are important as vehicles into a particular world. This being so, it is essential to know the worthiness of the models one considers emulating. Questions arise: Why follow a particular paradigmatic individual? What is the special nature of that paradigmatic individual that makes him or her worthy of being imitated? Answers to these questions are only hinted at in the works of Rūpa Gosvāmin.[40] We must turn to the more philosophically developed works of his nephew, Jīva Gosvāmin, to understand the early Gauḍīya Vaiṣṇava view on this important issue, for it is Jīva in particular who developed the concept of the *śakti*s or divine energies of Kṛṣṇa, which is the key to understanding the position of the Gauḍīya Vaiṣṇava paradigmatic individuals.

In his *Bhagavat Sandarbha,* Jīva uses two adjectives to describe the *śakti*s that emanate from Kṛṣṇa: *acintya* and *svābhāvika.*[41] The first term denotes the inconceivable nature of the *śakti*s. It brings to mind the longer philosophical call word of the Gauḍīya Vaiṣṇava Sampradāya—*acintyabhedābheda,* meaning here that the *śakti*s, as the manifest energies of the divine, are both different and not different from the source of their energy (*śaktimat*), Bhagavān, Lord Kṛṣṇa. The second term indicates that the *śakti*s are not exterior to the Lord, but are intimate and natural to him. In their totality they constitute his very self or essence, although, being infinite, Lord Kṛṣṇa transcends them all.

Jīva maintains in the *Bhagavat Sandarbha* that the entire universe consists of four aspects: the Lord himself (*svarūpa*), the great majesty or splendor of his form (*tad-rūpa-vaibhava*), the individual conscious souls (*jīva*s), and unconscious matter (*pradhāna*).[42] Each of these is an expression of a particular *śakti* of Bhagavān Kṛṣṇa. On the authority of a verse in the *Viṣṇu Purāṇa,* Jīva declares these *śakti*s to be threefold: the highest (*parā*), the knower of the field/body (*kṣetrajñā*), and ignorance (*avidyā*).[43] Jīva renames these three respectively: internal (*antaraṅgā*), marginal (*taṭasthā*), and external (*bahiraṅgā*).[44] An examination of each will be fruitful.

The internal *śakti,* also called the *svarūpa-śakti,* constitutes the intrinsic, essential, and perfect selfhood of Bhagavān and is therefore completely inseparable from him. It expresses itself as two of the four universal aspects mentioned above: the true essential nature of Bhagavān as he is in himself (*svarūpa*), and his great splendor (*tad-rūpa-vaibhava*), comprised of the parts and functions of his true self in the form of his eternal associates (*pārṣada*), residence (*dhāma*), and perfected devotees (*bhakta*s). More needs to be said concerning this important energy, but first we must examine the two remaining *śakti*s.

The external *śakti* is frequently called *māyā-śakti* by Jīva. In his *Paramātma Sandarbha,* Jīva mentions that *māyā* (typically translated as "illusion") has two functions: it conceals and distorts the essential nature (*svarūpa*) of Ultimate Reality.[45] Thus, though *māyā-śakti* is a manifestation of Bhagavān Kṛṣṇa in a partial form (called Paramātma), it is not "essentially" part of Bhagavān. It is external to the true nature of Bhagavān; it is reality misapprehended or inverted. *Māyā-śakti* is in opposition to the true nature of Bhagavān (*svarūpa-śakti*); the two cannot exist side by side. If the true nature of Bhagavān is perceived, then *māyā* is not perceived; if *māyā* is perceived, then the true nature of Bhagavān cannot be perceived. Jīva further explains in his *Paramātma Sandarbha* that *māyā-śakti* has two functions: *jīva-māyā* and *guṇa-māyā.*[46] Functioning as *jīva-māyā,* it blinds the true vision of the individual soul (*jīva*) by obscuring the soul's pure consciousness; thus the soul forgets its own true nature. As *guṇa-māyā,* it brings into existence the phenomenal world of matter (*pradhāna*), which is made up of the three *guṇa*s (aspects of ordinary existence); it sustains that world by balancing the three *guṇa*s and, then, by disturbing the three *guṇa*s, finally it changes or destroys the phenomenal world.

The last of the universal aspects, the individual conscious soul or *jīva,* is an expression of the marginal *śakti,* sometimes called the *jīva-śakti.* It is called marginal (*taṭasthā,* literally, "standing on the borderline") because it occupies the ambiguous position of being on the threshold of the other two *śakti*s, though it is distinct from both. The *jīva* does not belong specifically to either the essential *svarūpa-śakti* or the illusory *māyā-śakti,* but it is liable to the influence of both, and therefore displays a dual capacity. Sudhindra Chakravarti explains it this way: "Standing on the marginal line between Svarūpa-śakti and Māyā-śakti of Brahman, jīva-śakti (otherwise known as Taṭasthā-śakti) reveals a dual inclination for divine as well as mundane life."[47] The arrangement of the *śakti*s can be diagrammed as in Figure 1.

The human predicament as seen by the early Gauḍīya Vaiṣṇavas now becomes clear. *Māyā,* besides causing the appearance of the phenomenal world, ensnares the human souls (*jīva*s) with its allurements, causing them to forget their true nature and thereby mistakenly to identify them-

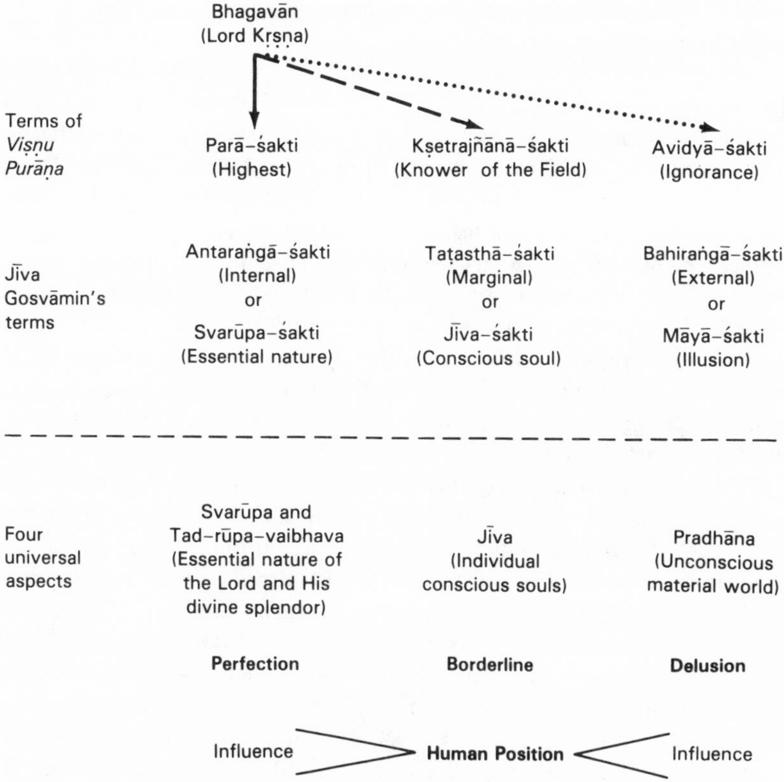

FIGURE 1. The Three Śaktis of Kṛṣṇa

selves with the phenomenal world, whereas the true identity of the souls is held to be a part (*aṃśa*) of Bhagavān. Under the illusory influence of the *māyā-śakti,* the soul cannot, until released, be truly connected with the essential nature of Bhagavān (*svarūpa-śakti*), which is by definition beyond the grasp of *māyā.* Tricked by the charms of *māyā,* the soul is doomed to be estranged from its true nature as a part of Bhagavān. Yet this ultimate affinity with Bhagavān does hold a ray of hope, even for the souls entrapped in the bondage of *māyā.* No souls are doomed forever. On the one hand, the illusory world of appearance caused by *māyā,* though "real," is not permanent. On the other, the souls—ultimately sharing as they do the spiritual essence of Bhagavān—never forfeit their inherent capacity to extricate themselves from the clutches of *māyā.* The subjugation of the soul, therefore, is not final. When it is emancipated from *māyā* and regains its true identity, it no longer occupies the ambiguous position of the marginal *śakti,* but returns to the essential, internal *śakti* of Bhagavān.

By what means is this accomplished? Those who have studied Gauḍīya Vaiṣṇavism can readily supply the answer: *bhakti.* "In every *jīva* there is an element of the expression of the Hlādinī-śakti, called Bhakti, which, when resorted to, rouses it from its prolonged slumber in the magic snare of Māyā and enables it to shake off the snare."[48] However, since previous scholars did not approach this question with a methodological framework that focuses attention on the importance of religious role taking, they failed to perceive the essential function paradigmatic individuals serve in this process. For just how is it that *bhakti* is aroused and the ideal religious world entered? I will demonstrate that this is accomplished by means of a particular religious practice (*sādhana*) which is heavily dependent on what I have been calling exemplary models or paradigmatic individuals.

We are now ready to suggest why it is that the models discussed above are worthy of emulation. Two Sanskrit terms used to designate the Gauḍīya Vaiṣṇava paradigmatic individual, eternally associated with Kṛṣṇa in the cosmic drama, are *parikara* and *pārṣada* (literally, "companions"). Significantly, we read in the *Bhagavat Sandarbha* that the *pārṣadas* are parts of Kṛṣṇa's own self (*tad–aṅga–bhūta*) and are part of his essential nature (*svarūpa-bhūta*).[49] And there we have it! The *pārṣadas* clearly occupy the realm of perfection; they were never touched by the delusion of *māyā.* We see in Figure 1 that the *svarūpa-śakti* is comprised of two aspects: the Lord himself (*svarūpa*) and his great splendor (*tad–rūpa–vaibhava*). Jīva Gosvāmin explains that the *pārṣadas*, the eternal companions of Kṛṣṇa, occupy the latter position. Being part of the *svarūpa-śakti,* the essential nature of Kṛṣṇa, they are apt models for perfection. Due to their location in the world of perfection they display a perfect form of *bhakti* called Rāgātmikā Bhakti.[50] Imitation of this perfect form of *bhakti* rouses *bhakti* in the slumbering soul and restores the soul to its true identity by pulling it out of its ambiguous position in the marginal state and guiding it into the perfect world of the *svarūpa-śakti.*

5

Entering the Cosmic Drama

Our type of creativeness (i.e. True Acting) is the conception and birth of a new being—the person in the part.

CONSTANTIN STANISLAVSKI, *An Actor Prepares*

It should now be quite evident that dramatic theory is central to Gauḍīya Vaiṣṇavism. Ultimate Reality is conceived of as a drama, and religious experience is understood in terms of dramatics. The question we consider in this chapter is: How is it that the Gauḍīya Vaiṣṇava enters and participates in the cosmic drama of Kṛṣṇa and there experiences *bhakti-rasa?* One of the main tasks undertaken by Rūpa Gosvāmin was to devise a method whereby the *bhakta* could enter the ultimately meaningful world of the Kṛṣṇa-līlā. The solution he came up with is the Rāgānugā Bhakti Sādhana. It is now appropriate to proceed to the heart of our analysis and examine this religious practice as it was first presented by Rūpa in the *Bhaktirasāmṛtasindhu,* examining additional sources where useful. Later historical developments, as well as select features of the practice as it is performed in Vraja today, will be discussed in the next two chapters.

Bhakti Sādhana

An important introductory matter must be cleared up before we plunge into the analysis of the Rāgānugā Bhakti Sādhana: Many previous scholars deny the existence of *sādhana* within the *bhakti* traditions. If this were indeed the case, we could proceed no further in our discussion of *bhakti-sādhana.* Therefore, it is necessary to examine briefly the statements of these scholars and evaluate their validity.

Early scholars of comparative religion were greatly attracted to the study of Hindu *bhakti,* for they thought that of all the religions of the world it was the one most similar to Christianity. Yet, I will contend, it is because they viewed *bhakti* through lenses coated with expectations of finding something similar to their own view of Christianity that they missed much of the complexity of *bhakti,* particularly those aspects that concern a study of *sādhana.* Misleading notions of *bhakti* were accordingly propagated, many of which reign today. It is thus necessary to consider some early representative definitions of *bhakti,* and to expand these definitions where they are unjustifiably narrow in their evaluation of the involvement of *sādhana.*

Two scholars who were very influential in early studies of Hindu *bhakti* in the West are Rudolf Otto and Nathan Söderblom. In 1929, Rudolf Otto published a small book of essays entitled *Christianity and the Indian Religion of Grace.*[1] Otto begins this set of essays by asking whether Christianity has any rivals in the East. (A rival for Otto is presumably a religion that offers a way to salvation in a manner similar to Christianity.) From all the religions of the East, he selects Hindu *bhakti* as that which comes nearest to Christianity. He writes: "The similarities present here are so important that it is tempting to consider this religion, viewed from the outside, as a sort of duplicate on Indian soil of that religion which emerged from Palestine and which we call Christianity."[2] Such statements must cause us to wonder what it was that Otto saw in Hindu *bhakti* that compared so favorably to Christianity.[3] How did he define *bhakti?*

Otto remarks that "the spiritual inheritance which has come down to us Protestants through Luther and the Reformation is *the doctrine of salvation by faith alone, without one's own merit or work.*"[4] And it is precisely in these terms that he defines *bhakti.* Otto writes of *bhakti* as love, devotion, and faith, and claims that in *bhakti:* "Salvation is not attained as a reward of our own works, but as a gift of grace, by a saving power above."[5] Therefore, he concludes: "Here to speak of a *means to salvation* is out of the question."[6] Here and elsewhere Otto insists that there is no "means to salvation" or, to use the Hindu term, *sādhana* in the religious systems of Hindu *bhakti.* Because of the opinion expressed in such declarations, *bhakti-sādhana* went unnoticed by Otto and those he influenced.

Two years after Otto's essays on *bhakti* were published, Nathan Söderblom published *The Living God: Basal Forms of Personal Religion.*[7] One of the essays in this book is called "Religion as Devotion: Bhakti." Writing on *bhakti* in the *Bhagavad-gītā,* Sōderblom comments: "Warren Hasting was right in writing that of all known religions this comes nearest to Christianity."[8] To understand what he meant by this statement we must look to see how he defines *bhakti.* His definition is strikingly similar to Otto's. He maintains that *bhakti* is "a new path of salvation which does not consist in works, offerings, or the exploits of ascesis, nor in knowledge and insight, but in faith, devotion, love towards a living personal deity or savior. . . . Salvation comes not by *tapas* and *yoga,* but by divine grace, which is *unmerited, not acquired, but given.*"[9]

To understand more fully Söderblom's presentation of *bhakti* as a type of religion quite different from and opposed to *yoga,* we must turn to another chapter in his book, entitled "Religion as Method: Yoga." In this chapter he explores the various methods or *sādhana*s of Hindu religion, particularly contemplation and meditation. According to Söderblom, *bhakti* has nothing to do with such methods. He writes: "Man cannot by any method or exercise stir up within himself faith. . . . *Bhakti,* faith, is no deed, for it does not depend on any exertion."[10]

There we have it: a definition of *bhakti* as love, devotion, and faith, not acquired by any methods (*sādhana*) or techniques, such as yogic meditation, but given by unmerited divine grace. Otto and Söderblom's views on *bhakti* remain very much with us today; perusing a number of recent volumes on Hinduism, I found this definition quite common. If, however, what Otto, Söderblom, and others have said is true, any attempt to analyze *bhakti-sādhana* is problematic; our subject disappears before we have the chance to examine it. Let us set aside for the moment what these scholars have said and ask ourselves this question: Do we find any form of *sādhana*—such as contemplation, meditation, hard work, discipline, and voluntary technical training—in the religious systems of Hindu *bhakti*? I strongly contend that this question must be answered in the affirmative.[11]

Let us consider for ourselves the very text on which both Otto and Söderblom base their arguments: the *Bhagavad-gītā.* Time and again Kṛṣṇa says in the *Bhagavad-gītā* that his *bhakta* is the one who constantly bears him in mind. The Sanskrit term used here is *smaraṇa* or *anusmaraṇa*—"bearing in mind," "concentrating on," or simply "remembering." For example,

> I am easily obtained, O Pārtha, by the ever-disciplined one who constantly bears me in mind (*smarati*), thinking of nothing else. (8.14)

Bhakti is equated again and again in the *Bhagavad-gītā* with *smṛti,* "remembrance," "concentration," or "contemplation." When attention is focused on this equation, *bhakti* appears as a technique of contemplation. How far we now are from Otto and Söderblom's definition of *bhakti!*

The contention that *bhakti* does indeed involve contemplative techniques finds additional support in the definitions of the prominent *bhakti* theorists themselves. In their works, one frequently finds *bhakti* defined in terms of *smaraṇa.* The famous *bhakti* theologian Rāmānuja, for example, defines *bhakti* as a constant bearing in mind, or remembrance (*dhrūvānusmṛti*).[12] I would therefore argue, and hope to demonstrate here, that *bhakti* has much to do with methodical effort and contemplative techniques. *Bhakti,* as *sādhana,* can perhaps best be defined as a specific contemplative technique which utilizes the emotions and involves concentration on and visualization of a deity. It is not as necessary to reject the previous definitions of *bhakti* as unmerited love, devotion, and faith (these elements are truly present in *bhakti*)[13] as it is to expand our definition to include contemplative techniques and other methods of *sādhana.* Rūpa Gosvāmin's own definition of *bhakti* makes this clear.

The term Rūpa uses to define *bhakti* is *anuśīlana* (BRS 1.1.11). Rūpa's nephew, Jīva Gosvāmin, clarifies the meaning of this rather ambiguous term in his commentary on Rūpa's definition. The common meaning of the term *anuśīlana* is "constant meditation" or "repeated practice." Jīva indicates, however, that it is derived from two distinct Sanskrit roots and is meant to encompass both. He glosses the first as *ceṣṭa,* or "effort"—that is, the actions or techniques by which Kṛṣṇa is obtained. The second gloss Jīva provides for the term *anuśīlana* is *bhāva,* the emotional content so important to *bhakti. Bhakti* is thus considered a contemplative technique that both utilizes and generates emotion. However, concerning the emotional aspect, which is crucial to the salvational process of *bhakti,* Rūpa comments:

> This emotion (*bhāva*) is born in two ways: either from the practice of *sādhana* or, for the extremely fortunate, by the grace (*prasāda*) of Kṛṣṇa and his *bhakta*s. But the first is common and the second is rare. (BRS 1.3.6)

Once again we have a clear statement from a prominent theorist of *bhakti* giving utmost importance to *sādhana.* Many Western scholars, such as Otto and Söderblom, must therefore be taken to task for their claim that methods of salvation have no place in Hindu *bhakti.* These scholars failed to understand that *bhakti* is both means (*sādhana*) and end (*sādhya*).[14] Keeping this in mind we are free to proceed with our examination of the Rāgānugā Bhakti Sādhana of Rūpa Gosvāmin.

Rāgānugā Bhakti Sādhana

A large section of the *Bhaktirasāmṛtasindhu* is entitled "Sādhana Bhakti." This section of the text is crucial to our investigation of the method devised by Rūpa to enable the Gauḍīya Vaiṣṇava to enter and participate in the cosmic drama of the Vraja-līlā. *Sādhana* is herein defined as the means by which an emotional relationship (*bhāva,* meaning here the foundational *sthāyi-bhāva*) is realized (BRS 1.2.2.). "Therefore," Rūpa advises, "the mind should be fastened on Kṛṣṇa by means of some method (*upāya*)" (BRS 1.2.4).

Many religious traditions share the view that we humans exist in a "lost state" (fallen, ignorant) and cannot reach the world of ultimate reality by means of our own inclinations, which are distorted by our fallen condition. Thus, we need the guidance of religious institutions to lead us to the desired goal. Two institutions are recognized within Gauḍīya Vaiṣṇavism as being valid for this purpose. The practitioner (*sādhaka*) is to follow either the injunctions of scriptural commands (*vidhi*) or the paradigmatic individuals (the various characters of the Vraja-līlā are collectively called the Vrajaloka). The first strategy is called Vaidhī Bhakti Sādhana, the second Rāgānugā Bhakti Sādhana. The two are traditionally considered sequential: Vaidhī Bhakti Sādhana prepares the way for the more esoteric path of Rāgānugā.

Vaidhī Bhakti Sādhana is defined by Rūpa in this manner:

> Bhakti is designated Vaidhī where it is made manifest by following the injunctions of scripture, and not from the attainment of a passion (*rāga*). (BRS 1.2.6)

Vaidhī Bhakti Sādhana is motivated by fear of sin[15] and is guided by sixty-four scriptural commandments that have been culled by Rūpa from a variety of Vaiṣṇava scriptures. In Vaidhī Bhakti Sādhana, the *bhakta* is called upon to perform various acts of service and worship to Kṛṣṇa. Here the *bhakta* is still acting within his or her ordinary self-identity, but begins to surrender that self through these acts of service. More important, through acts of Vaidhī Bhakti, both private and communal, the *bhakta* becomes familiar with the Vaiṣṇava scriptures, especially the stories of Kṛṣṇa and his intimates in Vraja. The Vaidhī Bhakti practitioner is required to frequently listen to (*caritra-śravaṇa*), remember (*smaraṇa*), sing about (*līlā-kīrtana*), meditate on (*krīḍa-dhyāna*), and otherwise celebrate the stories of the Vraja-līlā.[16] These exercises familiarize the practitioner with the world of Vraja and the characters who inhabit it. Staged dramas (such as the *rāsa-līlās*[17]) are also a favorite means of making the world of the Vraja-līlā objectively available to the interested community. Here the *bhakta* as spectator is presented with a vivid expression of the ultimate world of Vraja, which can provide an aesthetic foretaste of and powerful incentive for further pursuit of that world. The practitioner of Vaidhī Bhakti, continually studying the Vaiṣṇava scriptures that narrate the Vraja-līlā, moreover, is like an actor learning the script of a drama to be enacted. Unlike the "unwritten libretto" which guides the process of primary socialization,[18] here the libretto is objectively expressed and quite available in Vaiṣṇava scripture, such as the *Bhāgavata Purāṇa.*

The practitioner is to continue on the path of Vaidhī Bhakti, learning the stories of Kṛṣṇa and his close companions, until there arises the desire (*lobha;* internal desire, as opposed to the external law of Vaidhī, is the primary motivation for Rāgānugā) to assume one of the roles of one of Kṛṣṇa's paradigmatic companions, the Vrajaloka (BRS 1.2.291). That is, eligibility for Rāgānugā Bhakti Sādhana requires a desire to take on or enact one of the emotional roles in the script of the Vraja-līlā, thereby participating in the world of those events. When this occurs, the *bhakta* can begin to engage in the more esoteric Rāgānugā Bhakti Sādhana, the second stage of the *sādhana.*

The move onto the stage of Rāgānugā is necessary for further progress, because Vaidhī is judged to be inadequate for transporting one into the world of the Vraja-līlā. Only the Rāgānugā Bhakti Sādhana can accomplish this; it is therefore superior to the path of injunctions. "All the world has *vaidhī-bhakti* toward me, but through *vaidhī-bhakti* (alone) no one can gain the *bhāva* of Vraja."[19] Vaidhī Bhakti lacks the one ingredient essential to the Rāgānugā Bhakti Sādhana that makes entrance into the Vraja-līlā possible: the transformation of identity accomplished by assuming one of the exemplary emotional roles displayed by the original characters of Vraja (*Vrajaloka-anusāra*). We are told in the *Caitanya-caritāmṛta* that even the Vedas and Upaniṣads gave up their own regulations to become *gopīs* in Vraja.[20]

My introductory remarks should lead us to anticipate that any method devised to embody a new reality would necessarily involve some means of identity transformation. This is exactly what we do find in the Rāgānugā Bhakti Sādhana and, as might now be expected, the transformative technique is formulated in terms of dramatics. More precisely, the methods of the actor are conceptually employed by Rūpa in his effort to delineate the means by which the Gauḍīya Vaiṣṇava is to participate in the religious life.

The affectivity of the actor's art has been most aptly explored and expressed in the West by Constantin Stanislavski. Stanislavski discovered an intimate connection between internal emotions and external actions in a human being, between the psychological and the physical. He defined the subconscious as the internal world of uncontrolled emotions, and the conscious as the external world of controlled actions. Arguing in opposition to Freud, who stressed that the subconscious influences the conscious, Stanislavski asserted, with the support of his friend Ivan Pavlov, that the conscious can influence the subconscious.[21] This is an extremely important point for Stanislavski, since he believed the actor's success depended on being able to "turn on" the subconscious inner emotions in such a way that they would be in harmony with or similar to those of the character being portrayed. Stanislavski contended that access to the inner world of the subconscious could be gained through the external world of actions; thus the subconscious is dependent on the conscious. Action, he claimed, engages emotion. "Thus physical actions are the 'key' that lets the actor penetrate the inner world of the character portrayed."[22]

A similar notion seems to motivate much Hindu *sādhana*. Physical acts have certain positive effects on the inner mental state. One modern interpreter of the yogic tradition puts it this way: "The relationship between mind and body is complete and so subtle that it is no wonder that certain physical training will induce certain mental transformations."[23]

Stanislavski further maintained that the inner world of emotions is physically expressed. This also was supported by the leading Russian scientists of his day. "The thesis of Stanislavski that the complex of a human's psychological life—moods, desires, feelings, intentions, ambitions, for example—is expressed through simple physical actions has been confirmed by such scientists as Pavlov and I. M. Sechenov."[24] Thus, our access to the inner world of another, Stanislavski tells us, is through the other's external physical expression; it is with our bodies that we transmit to others our inner experience. The physical expression of an inner experience is what Bharata called the *anubhāva*. We shall soon see the importance of the physical expressions, or *anubhāva*s, of the paradigmatic individuals to the practitioner who hopes to gain access to their inner world.

Stanislavski informed his actors that the inner life of a character is approached by imitating the physical acts of that character. The anger of a particular character, for example, is realized by imitating the physical expressions of that particular character's anger—e.g., pounding fists on the table. This he called the "Method of Physical Actions."

The goal of such imitative action is to "live the role," that is, to completely enter into the world of the character one is imitating. "The very best that can happen is to have the actor completely carried away by the play. . . . he lives the part."[25] Stanislavski believed that the actor, through perfected acting technique, could totally identify with the character being portrayed, and for the time of the performance, could "become" that character and experience his or her world.[26] This is the experience he called "re-incarnation."

> An actor achieves re-incarnation when he achieves the truthful behavior of the character, when his actions are interwoven with words and thoughts, when he has searched for all the necessary traits of a given character, when he surrounds himself with its given circumstances and becomes so accustomed to them that he does not know where his own personality leaves off and that of the character begins.[27]

The whole of Stanislavski's research demonstrates that the actor can actually experience the inner world of a character (present only through a literary script) by imitating that character's physical actions (as expressed in the literary script). The assertion is that inner experience is reproducible, and that it is accessible through physical actions. Stanislavski considered the experience of re-incarnation the height of the actor's art, and argued that this experience has a tremendous effect on the life of an actor.[28] We would not be far from the mark to speak of Stanislavski's "re-incarnation" as a temporary transformation of identity. Accordingly, Stanislavski's insights lend themselves to an understanding of how dramatics might be used in an intentional process of identity transformation.

The generative notion behind the Rāgānugā Bhakti Sādhana is the realization that to obtain a goal one should imitate one who has already attained that goal. As Aristotle has said: To attain the good, find a good man and follow him.[29] Ideals of perfection are useless in the abstract; they must be embodied in human models to be useful in any transformative process. The "good man" for the Gauḍīya Vaiṣṇavas is one of the paradigmatic individuals of the Vraja-līlā. One of their exemplary roles (*bhāva*), Rūpa explains, must be taken on in order to enter and participate in the world of the Vraja-līlā. Viśvanātha Cakravartin, one of the most important commentators on Rūpa's works, expresses the notion this way. He says if a man wants milk he must go

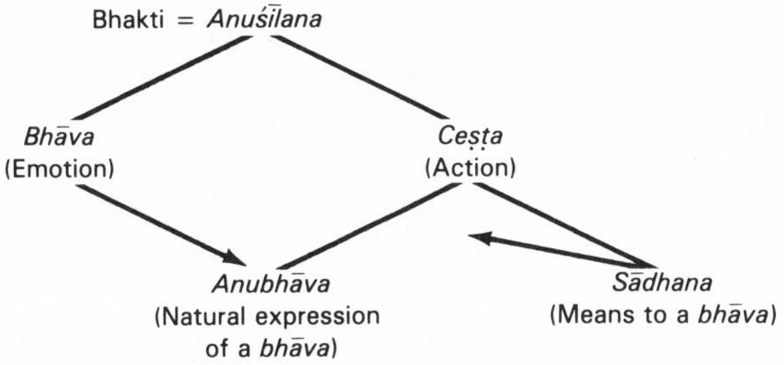

FIGURE 2. The relationship of *Anubhāva* and *Sādhana*

to a man who has milk and learn from him the process of tending cows and getting milk. That is, to obtain milk one should imitate one who already has milk. Likewise, to obtain the highest goal—perfect love (*prema*)—one must imitate and learn from those who have it, namely, the Vrajaloka, the original characters of Kṛṣṇa's playful drama in Vraja.[30]

The Rāgānugā Bhakti Sādhana is a ritual process formulated in terms of aesthetic experience. The key to understanding this *sādhana,* then, is an examination of the dramatic terminology used to express it. As previously mentioned, Rūpa defined *bhakti* with the term *anuśīlana* (BRS 1.1.11), and Jīva explained in his commentary that the intended meaning of the term included both emotion (*bhāva*) and action (*ceṣṭa*). In a later statement that I hold to be absolutely crucial for understanding the Rāgānugā Bhakti Sādhana, Jīva indicates that there are two sides to action: *anubhāva* and *sādhana.*[31] Thus the same act can be either an *anubhāva* or an act of *sādhana,* depending on its motive. If it is a spontaneous and natural expression of an inner emotion, it is an *anubhāva;* if it is an intentional act designed to acquire the inner emotion, it is *sādhana.* However, the two (*anubhāva* and *sādhana*) take the same physical form. The diagram of Figure 2 may be useful in visualizing the relationship of the *anubhāva* and *sādhana.*

Paralleling Stanislavski, Rūpa realized that the inner perfected emotions of the paradigmatic Vrajaloka were physically expressed as *anubhāva*s and were recorded in Vaiṣṇava scriptures, such as the *Bhāgavata Purāṇa.* In further agreement with Stanislavski's notions, he asserted that the only door of access to the inner emotions of the Vrajaloka is through these physical expressions or *anubhāva*s. Hence Rūpa's key insight: If the *bhakta* could somehow take on or imitate the *anubhāva*s of one of the exemplary Vrajaloka, he could obtain the salvific emotions of that character and come to inhabit the world in which that character resides—Vraja. This imitation of the ways of the Vrajaloka is the Rāgānugā Bhakti Sādhana.[32]

It is important to add, as Stanislavski also points out, that circumstances shape an emotion; thus the actors must know the circumstances of the emotional roles they are enacting. Likewise, Rūpa would have the *bhakta*s not only follow the *anubhāva*s of a particular character of Vraja, but also surround themselves with the given circumstances (*vibhāva*s) of the emotional role of that Vrajaloka. In this manner, the entire set of dramatic components of a particular Vrajaloka—*vibhāva, anubhāva, sāttvika,* and *vyabhicāryin* —are to be known and incorporated into the imitative activity.

Let us now carefully examine Rūpa's own technical definition of the Rāgānugā Bhakti Sādhana.[33] Rūpa first defines Rāgānugā as "that (method of *bhakti*) which imitates the Rāgātmikā (*bhakti*) clearly manifest in the inhabitants of Vraja" (BRS 1.2.270). Here the people of Vraja, the Vrajaloka, are the exemplary models, or what I have technically been calling the paradigmatic individuals. In their original interaction with Kṛṣṇa they display a perfect form of *bhakti* called Rāgātmikā Bhakti. This perfect *bhakti* forms the model which Rāgānugā Bhakti imitates. What the divine paradigmatic individuals naturally express, we humans must pursue with greater effort.[34]

Rūpa defines the exemplary Rāgātmikā Bhakti in terms of passion (*rāga*). Passion, he claims, is the "highest access" to the beloved Kṛṣṇa (BRS 1.2.271). In his *Bhakti Sandarbha,* Jīva Gosvāmin elaborates the meaning of this term and explains that passion "is that love which consists of an immense desire of a subject for union with the object of its desire."[35] It is that emotional force of attraction which binds together two separate entities. This after all is *bhakti,* a religious system which greatly values the tremendous power of emotions. Intense emotions, such as passion, produce an intense concentration on the object of the emotions, drawing subject and object together: thus they can bring about the desired goal of the mystic—union with the Divine. Passion, therefore, is held to be a highly valuable instrument for spiritual endeavors. Rāgātmikā Bhakti is defined by Rūpa as "that *bhakti* which is completely absorbed in or identical with that passion (*rāga*)" (BRS 1.2.272).

Since Rāgātmikā Bhakti was declared the model for the Rāgānugā Bhakti Sādhana, it had to be analyzed in detail. Having studied the Vaiṣṇava scriptures, Rūpa analyzed the exemplary models and their concomitant relationships conductive to *bhakti.* These he divided into two main categories: Amorous Bhakti (*kāma-rūpā*) and Relational Bhakti (*sambandha-rūpā*). Emotional relationships based on fear or hatred, though they are recognized as effective means of concentration, are rejected because of their "unfavorable" nature and therefore do not find a place among either Amorous or Relational Bhakti. They are rejected as proper models (BRS 1.2.276). The amorous emotion (*madhura-bhāva*) is always placed in a separate category because of its particular intensity.[36] It contains the essence of all other emotions, and then some.[37] Amorous Bhakti, says Rūpa, is perfectly represented by the *gopīs* of Vraja (BRS 1.2.284), and Relational Bhakti is best exemplified by the cowherders who thought themselves to be a relation of Kṛṣṇa—his parents, friends, and so forth (BRS 1.2.288). The female lovers of Kṛṣṇa belong to the category of Amorous Bhakti; all of the remaining paradigmatic individuals outlined in Chapter 4 belong to the category of Relational Bhakti. The true nature of Amorous and Relational Bhakti is revealed as perfect love (*prema*), the highest level of emotion—*bhakti-rasa* (BRS 1.2.289). Therefore, the Rāgātmikā *bhaktas*—the *gopīs*, Kṛṣṇa's mother Yaśodā and father Nanda, his friend Subala, etc.—who fully loved Kṛṣṇa in the original Vraja-līlā embody and objectively display perfect *bhakti* and, as such, become the exemplary models for the Rāgānugā Bhakti Sādhana.[38]

Rāgānugā, being modeled on Rāgātmikā, is also divided into two categories: Imitation of Amorous Bhakti (*kāmānugā*), and Imitation of Relational Bhakti (*sambandhānugā*); BRS 1.2.290). Imitation of Amorous Bhakti is modeled on the *gopīs* and itself is divided into two types, representing two distinct options within the *sādhana*. One consists of the desire for direct enjoyment and sexual union with Kṛṣṇa (*sambhogecchā-mayī*), and the other consists of the desire to share in the emotions of a superior Vrajaloka, usually Rādhā (*tattad-bhāvecchātmikā*) (BRS 1.2.298).[39] The second category of Rāgānugā, Imitation of Relational Bhakti, is modeled on the other remaining characters of the Vraja-līlā, such as Kṛṣṇa's parents, friends, and servants. Rūpa wrote three brief but important verses describing the Rāgānugā Bhakti Sādhana in the Relational mode.

305. Imitation of Relational Bhakti is declared by the sages to be that *bhakti* which consists of meditating on a relationship (with Kṛṣṇa)—fatherhood, and so forth—and imposing (*āropaṇa*) such a relationship on one's own self.

306. This *bhakti* is to be practiced by those practitioners desirous of parenthood, friendship, and so forth, by means of the emotions,

actions, and postures of the king of Vraja (Nanda), Subala, and other appropriate models.

307. It is written in the scriptures that a certain wise old man who lived in Kurupurī became perfected worshipping by Nārada's instruction the image of the son of Nanda (i.e. Kṛṣṇa) as his own son.

Rūpa explains that the Rāgātmikā *bhakta*s of Relational Bhakti, who are the perfect models for the Rāgānugā *bhakta*s following them, naturally possess a relationship with Kṛṣṇa that involves a particular identity (the Sanskrit term used for this identity is *abhimāna;* BRS 1.2.288). Jīva Gosvāmin further elaborates the connection between these particular relationships and identities in his *Prīti Sandarbha.*[40] He explains that the emotional relationship of a servant or younger relative (*prīti-bhāva*) involves the identity of oneself as being under the protection of Kṛṣṇa (*anugrāhābhimāna*); the relationship of friendship (*sākhya-bhāva*) involves the identity of oneself as a friend of Kṛṣṇa (*mitrābhimāna*); the relationship of the elder (*vātsalya-bhāva*) involves the identity of oneself as a guardian of Kṛṣṇa (*anugrāhakābhimāna*); and the amorous relationship (*madhura* or *priyatā-bhāva*) involves the identity of oneself as a beloved of Kṛṣṇa (*priyābhimāna*). One of these identities is an innate feature of the model Vrajaloka, since each of the Vrajaloka is considered to be eternally perfected (*nitya-siddha*). The Vrajaloka, therefore, do not require *sādhana.* Human beings, however, being influenced by *māyā,* which distorts and conceals their true identity, need to impose (*āropaṇa*) onto themselves through *sādhana* one of these ultimate identities which connects one to Kṛṣṇa in his world of Vraja.

Verse 306 quoted above states that the imposition of a true identity is to be accomplished by somehow taking on or imitating the emotions, actions, and postures of the paradigmatic individual in a particular emotional role. Mukundadāsa Gosvāmin points out, in his commentary on this verse, that these exemplary expressions of the Vrajaloka include the *anubhāva*s the practitioner must assume in the *sādhana.*[41] Hence we return to the insight shared in kind by Stanislavski; the inner emotional world of the Vrajaloka is reached by taking on their physical expressions (*anubhāva*s). Verse 307 informs us that one can become perfect (*siddha*) through this imitative action. In this case, however, perfection is accomplished by means of religious practice (*sādhana-siddha* or *āropaṇa-siddha*); it is not a natural and eternal perfection (*nitya-siddha*), a position strictly and uniquely reserved for the original Vrajaloka.

Perfection, or salvation, is here understood as total absorption (*āveśa*) in an eternal body which resides in Vraja. This body is called the "perfected body" (*siddha-rūpa* or the *siddha-deha*). The body (*rūpa* or *deha*) is understood in Gauḍīya Vaiṣṇavism to be the house of the soul or self (*ātman*). Identity is what locates the self in a particular body which resides in a particular world. To participate in the world of Vraja, one must occupy a body located in that world.[42] And to accomplish this one must develop an identity that connects one to such a body. A modern exponent of the Rāgānugā Bhakti Sādhana explains the process this way:

> By visualizing the Lord's *līlā* and performing mental acts in the identity (*abhimāna*) of the perfected body (*siddha-deha*) as instructed by the *guru*, occupation (*āveśa*) of the external illusory body will gradually be destroyed and occupation of the perfected body will occur.[43]

Salvation in Gauḍīya Vaiṣṇavism, then, can be seen as the shift of identity from the external illusory body to the true body (*siddha-rūpa* or *siddha-deha*) which is similar to the bodies of the exemplary Vrajaloka. The *Caitanya-caritāmṛta* reads: "One who worships by taking on the particular emotional role (*bhāva*) of an inhabitant of Vraja attains a body appropriate to that emotional role and thereby attains Kṛṣṇa."[44] This shift of identity is expressed in an interesting way in a seventeenth-century Sanskrit text which explicates the Rāgānugā Bhakti Sādhana in some detail.[45]

> Identity has three forms: dominant, medial, and slight. In the external state, the identity is slight in the *siddha-rūpa* and dominant in the ordinary body. In the half-external state, the identity is medial in the *siddha-rūpa* and medial in the ordinary body. In the inner state, the identity is dominant in the *siddha-rūpa* and slight in the ordinary body. In the pure inner state, the identity is totally in the *siddha-rūpa*. In the pure external state, the identity is totally in the ordinary body.[46]

This statement reflects the different levels of the *sādhana*. The external state refers to the ordinary, understood by Hindus as the illusory world of everyday. When a person's identity is predominantly in the ordinary body, that person inhabits the ordinary world. In contrast, the inner state refers to the interior meditative world of the ultimately real Vraja. The half-external state involves an awareness of oneself as belonging to both worlds. One scholar of modern Vṛndāvana suggested examples of each of these states in terms of worshipping the deity.[47] He explained that in the external state, the practitioner stands before a physical image (*mūrti*) and performs worship with an awareness of himself as a male *brahmin*, for example. In the inner state, the practitioner meditatively approaches Kṛṣṇa directly (*sākṣāt*) and performs worship as a *gopī*. In the half-external state, the practitioner has a dual identity and performs both forms of worship at the same time. The practitioner progressively moves from the external state to the half-external state, and from the half-external state to the inner state. The pure inner state is the goal; it is a condition in which the identity totally occupies the *siddha-rūpa* and the practitioner is completely absorbed (*āveśa*) in the world of the Vraja-līlā.

How, then, is this transformation accomplished? Rūpa gives two essential and instructive verses for the practitioners of the Rāgānugā Bhakti Sādhana (BRS 1.2.294 and 295). The first of these instructs the practitioner to visualize the *līlā* of Kṛṣṇa in Vraja. That is, the practitioner is to make the *līlā* vividly present in the mind.[48] But the practitioner is not to stop at simply visualizing Kṛṣṇa's *līlā*. Rūpa further instructs the practitioner to strive to enter and participate in that *līlā* through acts of service or a loving performance (*sevā*) directed toward Kṛṣṇa.[49] He insists that "one desirous of attaining one of the emotional roles (*bhāva*s) of the Vrajaloka should do performative acts of service (*sevā*) in a manner which imitates or follows (*anusāra*) the original characters of Vraja with both the practitioner's body (*sādhaka-rūpa*) and the perfected body (*siddha-rūpa;* BRS 1.2.295)."[50] Some examples of service or performative acts by Gaudīya Vaiṣṇava adherents, with the practitioner's body and the perfected body, will be explored in Chapter 7. The *Rāgānugā-vivṛtti* amplifies the notion that imitative performative acts of service (*sevā*) accomplish what the Rāgānugā Bhakti Sādhana sets out to achieve.

> The establishment of the identity in the *siddha-rūpa* is accomplished by performative acts (*sevā*). The destruction of the false identity in the ordinary body is accomplished by performative acts. The first identity is due to the power of consciousness (part of the essential nature of Kṛṣṇa—*svarūpa-śakti*); the second is due to *māyā*. The first is to be accomplished; the second is to be destroyed.[51]

The goal of the Rāgānugā Bhakti Sādhana, then, is a shift from the "ordinary" world of everyday to the Ultimate Reality of the Vraja-līlā. Alfred Schutz and others had described the dynamics of such reality shifting as inexplicable, haphazard "leaps" or "slips." But I have suggested that the Gauḍīya Vaiṣṇavas accomplish this feat by the intentional formation of an identity which is intimately connected to the world of meaning expressed as Vraja. The assumption of a new identity is a crucial feature of the Rāgānugā Bhakti Sādhana. In fact, Jīva

Gosvāmin defines the practitioner of this *sādhana* as one who possesses one of the special identities of the original Vrajaloka.[52] The Gauḍīya Vaiṣṇavas achieve the world of Vraja by assuming through dramatic technique an identity which locates them in that world. There are, of course, degrees in this process. At first, the practitioners merely copy the character they are striving to become. But the successful practitioner's goal is to become so totally identified with a character in the Vraja-līlā that he or she *really is* that character. This total identification with and absorption in (*āveśa*) the world of the dramatic character is similar to what Stanislavski has called "re-incarnation."

The comparison of the Gauḍīya Vaiṣṇava concept of *āveśa* ("absorption in") and Stanislavski's concept of "re-incarnation" is justified, though it needs qualification. *Āveśa* is sometimes used within a dramatic context in Gauḍīya Vaiṣṇava texts. We read in both the *Caitanya–bhāgavata* and the *Caitanya–caritāmṛta* that on several occasions Caitanya or his close associates enacted plays of Vaiṣṇava stories. These plays had a very serious tone, for we are told that during these dramatic enactments some of those involved became completely absorbed in (*āveśa*) the characters they were enacting.[53] This type of serious dramatic acting, where the actors are totally carried away by their roles, must certainly have influenced the development of the Rāgānugā Bhakti Sādhana. Moreover, this experience of total absorption in a dramatic character is exactly what Stanislavski meant by the term re-incarnation.

There are, however, some major differences between the actor practicing the Stanislavski Method and the practitioner of the Rāgānugā Bhakti Sādhana aiming at complete absorption in a dramatic role which is understood to be his or her true identity in the eternal world of Vraja. These should be pointed out. While the actor temporarily engages in a vast variety of roles, the Rāgānugā practitioner strives to realize one role in one play for all time. Plays for the ordinary actor are the creation of a playwright; the only play the Rāgānugā practitioner considers worthy of enactment is Ultimate Reality itself—the Vraja-līlā, as revealed in Vaiṣṇava scripture. Furthermore, the ordinary actor does not completely forget he or she is an actor. To overcome the "sham" of acting as someone else, Stanislavski supplied his actors with the "magical if."[54] A dual consciousness is required of the actor. Actors never lose sight of their ordinary identities (social selves?) while acting "as if" they were someone else. Otherwise, they would never leave the stage. In great contrast, the goal of the practitioner of the Rāgānugā Bhakti Sādhana is to forget the old self completely and permanently, and become thoroughly absorbed in the true identity of the part—e.g., the *siddha-rūpa* as a *gopī.* The Rāgānugā practitioner's aim is never to leave the cosmic stage of the Vraja-līlā; imitation is to become permanent participation.

Previous studies of Gauḍīya Vaiṣṇavism have freqently included a brief discussion of the Rāgānugā Bhakti Sādhana. My own views are perhaps closest to those of David Kinsley, who remarks that "in the theory of *rāgānugā* the end is complete identification with a transcendent model."[55] Interestingly, Kinsley also compares the *bhakta*'s transformation of identity to the experience of the classical actor in Indian theatre. The actor in classical drama aims at achieving a state called the "other mind," which is remarkably similar to what Stanislavski has called re-incarnation. "And insofar as the actor completely identifies himself with the actions he is imitating, insofar as he may become overwhelmed or seized by his model, he comes to participate in the state of being called the 'other mind' (*anyamānas*)."[56] Kinsley goes on to compare explicitly the classical Indian actor and the *bhakta.*

> In Bengal Vaiṣṇava devotional theory a particularly successful means of experiencing *bhakti* is by employing the methods of the actor in classical dance-drama. The effectiveness of the actor's method derives from the fact that he must discipline himself to forget or control his own emotions and reactions in order to let the character he is portraying appear through him. By playacting the devotee seeks to lose his identity by accommodating the identity of another. The devotee, like the actor, seeks to play a role and, like the actor in classical dance-drama, seeks to play it well and effectively by so completely identifying with the role that he loses himself, or forgets himself.[57]

Kinsley's analysis agrees for the most part with my own interpretation of the Rāgānugā Bhakti Sādhana. I would, however, insist that the *bhakta* does not so much strive to "forget himself" as to discover his true self. Loss of identity is not the goal—one cannot meaningfully exist without an identity—rather, the goal is the discovery and realization of the true identity (*abhimāna*) which connects the *bhakta* to Kṛṣṇa and the world of Vraja. Moreover, the limitations of the comparison between the *bhakta*'s experience and that of a Stanislavski Method actor are equally applicable to Kinsley's comparison of the *bhakta* and the classical Indian actor.

Other scholars of Gauḍīya Vaiṣṇavism have offered brief interpretations of the Rāgānugā Bhakti Sādhana. Differences in interpretation have arisen, however, over two related issues. The first concerns the exact meaning of the term *anuga,* which appears in the Sanskrit compound *rāga-anuga* ("following" [*anuga*] the passion [*rāga*] exemplified by the Vrajaloka). The second concerns differing views about the appropriate position for a *bhakta* to assume in the effort to experience Kṛṣṇa's *līlā.*

Rūpa supplied the term *anusāra* as a synonym to define *anuga* in his discussion of Rāgānugā and seemed content to leave it at that (BRS 1.2.270 and 295). The term *anusāra,* however, perhaps most neutrally translated as "following," proved to be equally ambiguous. S. K. De has led the way in translating *anusāra* with the English word "imitation." He writes:

> One desirous of the way of realization will adopt the particular Bhāva (e.g. Rādhā-bhāva, Sakhī-bhāva, etc.) of the particular favourite of Kṛṣṇa according to his or her Līlā, Veśa and Svabhāva, and live in the ecstasy of that vicarious enjoyment. The emotion . . . is engendered by external effort, by elaborately *imitating* the action and feeling of those connected with Kṛṣṇa in Vraja.[58]

A number of other Indian scholars interpreting Gauḍīya Vaiṣṇavism for English readers present Rāgānugā as a process of "imitation." S. C. Cakravarti comments:

> The earth-bound souls, getting attached to the Lord in course of listening to His charming attributes and sportive actions, can only *imitate* some of the loving services of the eternal attendants in Vṛndāvana in accordance with their respective capacities and tastes. Such *imitation* is called Rāgānugā Bhakti, or a form of devotion that follows in the wake of Rāgātmikā Bhakti.
>
> Rāgānugā Bhakti too involves an external effort, since *imitation* of the action and feelings of Lord's attendants forms one of its elements.[59]

O. B. L. Kapoor explains:

> Rāgānugā-bhakti is only an *imitation* of Rāgātmikā-bhakti. In Rāgānugā-bhakti the devotee . . . *imitates* the particular mode of Rāgātmikā-bhakti that suits his natural inclination.[60]

And finally, N. C. Ghose, in a work that predates De, presents Rāgānugā in these terms:

> When the *jiva,* on hearing of the unbounded bliss which prevails in Brindaban amidst the female Kunja inmates there, is tempted to procure equal bliss for himself, he . . . tries to *imitate* the conduct of the Gopies of

> Brindaban, tries to become moulded in their sentiments. . . . This following the tread of the Gopies to get the turn of their hearts is named Raganuga Bhakti.[61]

Some scholars have objected to the use of the word imitation and would rather speak of *anuga* or *anusāra* as "putting oneself in the service of" or "becoming subordinate to" one of the Vrajaloka. This is particularly true of modern Bengali writers,[62] who point out that some important commentaries on Rūpa's works make much of the fact that Rūpa used the term *anusāra* as a synonym for *anuga,* not *anukāra.*[63] One finds that nearly every modern translation of and commentary on the *Bhakti-rasāmṛtasindhu* in Bengali and Hindi strongly make this assertion.[64] One is to "follow" (*anuga* or *anusāra*) the Vrajaloka, not "merely imitate" (*anukāra*) them. *Anuga,* they insist, is not *anukāra.*

In his *Bhakti Sandarbha,* however, Jīva Gosvāmin illustrates that *anuga* includes *anukāra.* That is, *anuga* is "imitation" plus something else. In the context of a discussion of Rāgānugā, Jīva says that because the demoness Pūtanā merely imitated (*anukāra*) a wet nurse for Kṛṣṇa (by suckling him at her breast, though her intention was to kill him) she attained salvation. He writes:

> Scripture tells that a state similar to the state of those who possess Rāgātmikā Bhakti may be obtained by means of mere imitation (*anukaraṇa*), even with a bad intention. There is the case of Pūtanā's imitation of a wet nurse. How much easier could this occur by conforming to a constant and complete *bhakti* which is identical to that of the Rāgātmikā models and shares in their longing.[65]

Jīva does not claim that Rāgānugā is altogether different from imitation (*anukāra*), but that it is something *more* than "mere imitation." That something more is the right intention, the intention of gaining the emotional state (*bhāva*) of the model one is imitating. Without this spiritual intention, imitation would be but mere impersonation.

Donna Wulff offers the translation "conforming (oneself) to" for the term *anusāra.*[66] I like this translation, especially considering the history of the use of the term *conformatio* by the monastic theologians, such as Bernard of Clairvaux, who were concerned with the imitation of Christ.[67] Yet I think "imitation" is an equally acceptable translation, if we keep in mind that it is not "mere imitation," but an imitation that includes a particular intention: the realization of the world of the one who is being imitated.[68] Those writers who would deny the imitative nature of Rāgānugā are, I think, speaking from a more recent and limited understanding of the *sādhana,* though of course imitation does certainly involve subordination to the one being imitated.

Consideration of the manner in which a practitioner of the Rāgānugā Bhakti Sādhana is to follow the Vrajaloka has an important relationship to the second interpretive issue—namely, the position the practitioner is to assume to participate in Kṛṣṇa's *līlā.* We might ask with Wulff: "To what extent is the devotee a spectator of the eternal *līlā* of Kṛṣṇa and his intimate associates in Vraja, and to what extent may he become a participant in this *līlā?* And if he becomes a participant, what role or roles does he assume?"[69] Two opinions emerge: one arguing for direct participation, the other for the more distanced role of a witness. Following the lead of De, Edward Dimock presents Rāgānugā as an imitative process which leads to "becoming." He writes:

> The bhakta devoted to Kṛṣṇa takes on himself one or another of these attitudes (*bhāvas*) according to his inclinations and capacity, and *becomes,* by a mental-training process, . . . one or another of the people in the *Bhāgavata* stories in his relation to the Lord.
>
> . . . The Vṛndāvana-līlā can actually be recreated in the life of the worshipper by his taking on the personality of one of Kṛṣṇa's intimates in Vṛndāvana. Constant thought, remembrance, reflection, and action lead to *becoming.*[70]

According to Dimock, Rāgānugā is the means by which the *bhakta* becomes a different participant in Kṛṣṇa's *līlā.*

Differing opinions exist, however. After surveying the works of both De and Dimock, Wulff comments:

> Other interpreters of the tradition have taken a rather different position. Drawing on the works of Bimanbehari Majumdar and others, Joseph O'Connell has argued in his thesis that the primary models whose feelings and actions toward the Lord are imitated by the devotee are not the close associates of Kṛṣṇa in Vraja, but the *mañjarīs* or maidservants who humbly serve Rādhā and the other prominent *gopīs.*[71]

The theory referred to in this quotation states that the practitioner of Rāgānugā aims at being more of a spectator/servant, called a *mañjarī,* of the *līlā* of Kṛṣṇa and his intimates than a direct participant in it. (Much more will be said about this *mañjarī* figure in the following chapter.) Wulff sets this interpretation against the one that presents Rāgānugā as a striving for direct participation in the *līlā* and maintains that the two interpretations are in direct conflict. Wulff finds further evidence for the distanced role in the writings of Shashibhusan Dasgupta, who writes:

> Śrī Caitanya placed himself in the position of Rādhā and longed with all the tormenting pangs of heart for union with his beloved Kṛṣṇa; but the Vaiṣṇava poets, headed by Jayadeva, Caṇḍīdāsa and Vidyāpati, placed themselves, rather in the position of the Sakhīs, or female companions of Rādhā and Kṛṣṇa, who did never long for their union with Kṛṣṇa,—but ever longed for the opportunity of witnessing from a distance the eternal love-making of Rādhā and Kṛṣṇa in the supranatural land of Vṛndāvana.[72]

Dasgupta is unfortunately referring to pre-Caitanya Vaiṣṇava poets; however, the same view can be found among scholars referring to post-Caitanya Gauḍīya Vaiṣṇavism. For example, in his study of Gauḍīya Vaiṣṇava doctrine, A. K. Majumdar maintains that a Rāgānugā practitioner must consider himself a servant of the people of Vraja and thereby witness their love play.[73] N. N. Law, in his informative article on the Rāgānugā Bhakti Sādhana, presents the role of the practitioner as that of a *mañjarī,* who desires to witness the ongoing *līlā* of Rādhā and Kṛṣṇa and to serve as an attendant.[74]

We then find two views—conflicting views, Wulff tells us—in the secondary literature dealing with the Rāgānugā Bhakti Sādhana. One presents the goal of Rāgānugā as direct participation in the *līlā;* the other presents the goal of Rāgānugā as a less direct witnessing of the *līlā.* Wulff concludes that the latter of the two is a later historical development, due in part to the presence of the British. She remarks: "It is difficult not to see in the practice of *mañjarīsādhana* a greater sense of distance between these characters and the ordinary human *bhakta* than Rūpa's *Vidagdhamādhava* or his theory seems to require."[75] While I agree that the less direct approach rose to importance through historical developments, I would go on to argue that the two scholastic views that Wulff brings to our attention are based on two different options available to the practitioner within the tradition itself (though this fact seems to have escaped many Western scholars), which were briefly established by Rūpa in the *Bhaktirasāmṛtasindhu.* A few Indian scholars writing in English have noticed these two options,[76] but the essential difference has not been adequately defined and appreciated. To understand the full range of the Rāgānugā Bhakti Sādhana we must examine these two options in detail. Only then can we attempt to answer the question of whether the *bhakta* is to be a spectator of or an actor in the *līlā,* and if an actor, what role is appropriate.

Two Options for the Sādhana

The two options available to a practitioner of the Rāgānugā Bhakti Sādhana are most clearly indicated by Rūpa in verses 298–300 of his discussion of Rāgānugā.[77] In these verses, Rūpa defines two paths for the Imitation of Amorous Bhakti (*kāmānugā*). The first consists of the desire for direct sexual enjoyment of Kṛṣṇa (*sambhogecchāmayī*); the second, the desire to share in the emotions of one of the *gopī*s (*tattadbhāvecchātmikā*). The goal of the first is said to be unmediated amorous involvement with Kṛṣṇa, while the goal of the second is the vicarious enjoyment of the emotional experience of one of the original *gopī*s of Vraja (almost exclusively Rādhā). Rūpa declares that the practitioner has these two options for realizing the emotions of Vraja.

Vṛndāvana scholars sometimes connect these verses in the *Bhaktirasāmṛtasindhu* with a few explanatory remarks in Jīva's *Prīti Sandarbha.* The connection is well justified, consistent with the commentaries, and aids in a deeper understanding of the two options under consideration. In the context of a discussion of the emotion of the female lovers of Kṛṣṇa (*kāntā-bhāva*), the highest of the goals to be attained, Jīva writes:

> And this emotional role (*bhāva*) is of two types: One consists of direct enjoyment (*sākṣād-upabhoga*); and the other consists of the vicarious enjoyment (*anumodana*) of it. The first is for the *nāyikā*s; the second is for the *sakhī*s. Each, however, participates in the other.[78]

Thus the two options for the Imitation of Amorous Bhakti are as follows: The Rāgānugā practitioner can develop through *sādhana* a direct (*sākṣāt*) relationship with Kṛṣṇa in the world of Vraja. In the amorous relationship, the one who plays this role is known as a heroine (*nāyikā*) or a group leader (*yūtheśvarī*). Alternatively, the practitioner can take a role that is, although less direct, potentially more intense. The idea here is to become a friend (*sakhī*) of one of the *nāyikā*s or *yūtheśvarī*s, preferably Rādhā, and thereby witness her love affairs as a supportive spectator. Notice, however, that even as a "spectator" the *bhakta* is required to assume a particular role in the drama; in this case the role is that of a female friend, a *sakhī,* who serves the divine couple and helps bring about their union. Perhaps it would be best to think of the *nāyikā* as the principal actress and the *sakhī*s as supporting actresses.

Kuñjabihārī Dāsa of Rādhākuṇḍa explains the two options this way:

> Sexual union/enjoyment (*sambhoga*) is to be spoken of thus: It is a special love which experiences directly the mingling of the limbs of the group leader (*yūtheśvarī*), such as Śrī Rādhā, for the pleasure of Śrī Kṛṣṇa. That *bhakti* which consists of the desire for the emotional role of the heroine (*nāyikā*) and is associated with this special love is called The Desire for Sexual Enjoyment (*sambhogecchāmayī*).
>
> And that *bhakti,* which consists of the desire to help bring about and vicariously enjoy (*anumodana*) the special mingling of the limbs of the group leader, such as Śrī Rādhā, with Śrī Kṛṣṇa, which manifests the special emotion of a female friend (*sakhī*) that draws together the heroine and the hero (i.e., Rādhā and Kṛṣṇa), which experiences its own special kind of pleasure, and which is exemplified by the *sakhī*s, such as Śrī Lalitā, Viśākhā, etc., and the *mañjarī*s, such as Śrī Rūpa, Rati, etc., is called The Desire to Share in One of Their Emotions (*tattadbhāvecchātmikā*).[79]

Apparently Rūpa does not give priority to either of these paths. Jīva, however, in his commentary on the relevant verses in the *Bhaktirasāmṛtasindhu,* devalues the first in comparison with the second. He remarks that the direct approach, *sambhogecchāmayī,* is to be understood as that which imitates the excessive desire for sexual enjoyment exemplified by Kubjā in verse 1.2.287 (*kāmaprāya-anugā*), and is therefore incomplete. At the same time, he declares that the indirect approach, *tattadbhāvecchātmikā,* is to be understood as the desire to share in the special emotions of one's favorite woman in Vraja and is therefore praised as the primary mode of the Imitation of Amorous Bhakti (*kāmānugā*). Jīva's word on this matter is accepted as authoritative by most later writers.[80] From these discussions two options for the Rāgānugā Bhakti Sādhana emerge, two options which can be, and indeed were, extended and applied to the other emotional roles available to the practitioner.

The spiritual quest in Gauḍīya Vaiṣṇavism can best be expressed by the terms *viṣaya* (the "object" of an emotion) and *āśraya* (the "vessel" of the emotion). The quest is accordingly defined as a search for the true *viṣaya* and the true *āśraya.* The true *viṣaya* is, of course, always Kṛṣṇa, though he assumes different forms appropriate to the various relationships. But the true *viṣaya* can only be discovered in conjunction with a simultaneous discovery of one's true *āśraya,* one's own true emotional vessel deep within, which experiences the divine. The *āśraya* within is intimately connected with the true identity and is conceived of as the *siddha-rūpa,* the eternal body in which one is to play a part in the eternal drama of Kṛṣṇa. The two options represent two different ways of locating and realizing one's true *āśraya,* one's religio-aesthetic "vessel."

In the first option, the direct approach, the practitioner strives to take the role of a direct participant in the activities of Kṛṣṇa's *līlā*—as friend, parent, or lover of Kṛṣṇa—and thereby generate the salvific *sthāyi-bhāva* of love for Kṛṣṇa, which leads to the experience of *bhakti-rasa*. That is, here the practitioner takes the position of a direct vessel (*āśraya*) of a particular exemplary emotion. This seems to be the option noticed and reported by Dimock and those who refer to the process of Rāgānugā as "becoming" one of the main characters in the *līlā*. A rather complicated qualification, however, was later added to this approach.

Where Rūpa is again silent, Jīva sounds a note of warning. In his commentary on the verses describing the Imitation of Relational Bhakti in the *Bhaktirasāmṛtasindhu*, Jīva points out that there are two possible ways of understanding the direct approach.[81] One way is to think of oneself as an independent (*svatantra*) character in the drama; the other is to think of oneself as completely non-different (*abheda*, when the call word of Gauḍīya Vaiṣṇavism is "difference-in-nondifference" [*acintya-bhedābheda*]) from a particular original character. The first way, for example, involves conceiving of oneself as a unique *gopī* (for every soul has a very particular form), whereas the second way involves conceiving of oneself as one of the original characters, such as Rādhā. Jīva condemns the second way as improper, since it confuses the distinction between a human soul and one of the original models, who have never been touched by the influence of *māyā* and are held to be part of the essential nature (*svarūpa-śakti*) of Kṛṣṇa himself. Viśvanātha Cakravartin adds that this involves an offence that he calls *ahaṃ-grahopāsana*, which means "taking oneself for the object of worship."[82] The divine model and the imitator never become completely identified. The distinction is subtle but important. Human souls can become like, but not identical to (*acintyabhedābheda*), the original Vrajaloka such as Rādhā. A child enters the world of the model parent by imitating and thus taking on an identity presented by the parent, but in the process does not exactly become the parent. Similarly, the Rāgānugā practitioners dramatically imitate and thereby *conform themselves to* a role presented by a particular original character of Vraja, but according to the later warnings of Jīva and Viśvanātha, must not then think they are that original character, but rather must become independent characters in the Vraja-līlā, similar to the original characters, the Vrajaloka. For example, the practitioner of this option can become an independent *gopī*, with a distinct name, and so forth, who has a direct relationship with Kṛṣṇa, but cannot become Rādhā. In fact everyone, we will see, has a unique "double" defined in terms of the perfected body (*siddha-rūpa*). Thus the practitioner, if successful, can become an inhabitant of Vraja, *a* Vrajaloka, but not one of *the* original Vrajaloka.

In the second option, the indirect approach, the practitioner strives to take a role which involves putting oneself in the close service of one's favorite inhabitant of Vraja, and experiencing the salvific emotions of that inhabitant vicariously. Here the *bhakta*'s favorite inhabitant of Vraja, most frequently Rādhā, is the vessel (*āśraya*) whose emotions are relished by the spectator-*bhakta*. Although the two options seem to have been presented as equally valid by Rūpa in the *Bhaktirasāmṛtasindhu*, this second option was soon judged to be superior. When I first learned of this fact, I had what I think is a typical Western response: I thought that the direct approach was rated lower because it put one in competition with Rādhā, who has steadily gained in popularity among Vaiṣṇavas. But this is not the most relevant reason. Again, the context is aesthetics. The intensity of an emotion is in proportion to the depth of its "vessel" (*āśraya*). In the first option, the direct approach, the *bhakta* must rely on the depth of his or her own vessel while experiencing the religious emotions of the Vraja-līlā. By contrast, in the second option, the indirect approach, the *bhakta* relies on the vessel of his or her chosen inhabitant of Vraja, usually Rādhā whose emotional vessel or *āśraya* by definition is infinitely deep. The *bhakta*s are fond of saying that Rādhā's emotion is like an ocean compared to the drop of another. To experience even a small portion of Rādhā's love—that would truly be bliss! The goal of this indirect approach is a total emotional identification with Rādhā. When this occurs, Rādhā's emotional vessel and the practitioner's emotional vessel merge and become one.

The practitioner of the indirect approach usually assumes the role of a girlfriend of Rādhā, a *sakhī*. Emotional identity with Rādhā is a characteristic of the *sakhī*s who are more devoted to Rādhā than to Kṛṣṇa. These are the *nitya-sakhī*s or *prāṇa-sakhī*s mentioned in Chapter 2.[83] The particular quality which allows a unity of emotion between friends is said to be "trust" (*viśrambha*). Rūpa defines this technical term as a special confidential communication (*gāḍha-viśvāsa-viśeṣa;* BRS 3.3.106). Jīva glosses its meaning as a deep trust, which produces a "mutual non-difference" in all things.[84] Commenting on these verses and their relevance to the emotional identity between a heroine (*nāyikā,* i.e., Rādhā) and her friend (*sakhī*), Kuñjabihārī Dāsa explains:

> As a result of this deep trust (*viśrambha*), the heart of the heroine (*nāyikā*) is known to the friend (*sakhī*). That is to say, even if the heroine

does not speak, the friend is able to understand the emotion in the heart of the heroine from the tiniest hint.[85]

The practitioner following the indirect approach strives to take the role of a *sakhī* (or, as we shall see in the following chapter, a servant of a *sakhī* called a *mañjarī*), and thereby to achieve emotional identity with Rādhā. In this emotional communion, the practitioner as *sakhī* shares in the *bhakti-rasa* Rādhā experiences in her relationship with Kṛṣṇa. The *sakhī*-role involves serving the main heroine Rādhā and trying to bring about her union with Kṛṣṇa. The *Caitanya–caritāmṛta* even tells us that the *līlā* of Rādhā and Kṛṣṇa could not take place without the aid of the *sakhī*s.[86] The practitioner of the indirect approach, through a relationship defined in terms of confidential trust (*viśrambha*), becomes "plugged in" to Rādhā's emotional experiences arising from her love affairs with Kṛṣṇa.[87] And since Rādhā by definition (being the *hlādinī-śakti*—the Joyful Power of Kṛṣṇa) has the capacity for the most intense emotions possible, this approach gives one a much greater emotional experience of love for Kṛṣṇa than the direct approach. Kṛṣṇadāsa Kavirāja, in the *Caitanya-caritāmṛta,* speaks of the position of the *sakhī*s in this way: "The nature of the *sakhī*s, is an inexplicable matter; the minds of the *sakhī*s are never preoccupied with their own love play with Kṛṣṇa. They help bring about the love play of Rādhā with Kṛṣṇa and thereby obtain pleasure millions of times greater than they would from their own direct love play."[88] This latter option, the indirect approach, is the sole focus of those interpretations which Wulff contrasts with Dimock's interpretation. Much of the interpretive battle over the Rāgānugā Bhakti Sādhana, then, is resolved when we realize that there are really two options for the *sādhana,* both of which are recognized by the tradition in literature and practice, although the second came more and more to be judged the superior.

The practitioner following the direct approach, the *nāyikā* option, first visualizes the *līlā* and views it as a spectator, but then moves onto the stage of the *līlā* as a direct actor-participant. The practitioner following the indirect approach, the *sakhī* option, must assume an actor-role which allows access to the secret *līlā* of Rādhā and Kṛṣṇa. The love *līlā* is then viewed and vicariously enjoyed from the perspective of a spectator or, perhaps more precisely, from the perspective of a supporting actress. The relevance of Jīva's statement in the *Prīti Sandarbha* that each option participates in the other thus comes to light. The *nāyikā* also sometimes serves as a supportive *sakhī,* and a *sakhī* sometimes directly participates as a *nāyikā.* The division between actor and audience is continually blurred in *bhakti* aesthetics.[89] Yet no matter which option the practitioner opts for (or better, which the *guru* assigns), all agree that the ultimate world of the Vraja-līlā cannot be experienced within the ordinary identity associated with the ordinary body. A transformation is required.

The Perfected Body: *Siddha-rūpa*

One of the most frequent statements I heard in my interviews with the Rāgānugā practitioners in Vraja was this: "You cannot participate in Kṛṇṣa's *līlā* with this body. You must first obtain a new body." Many stories relating to this notion circulate in Vraja. The mighty Hindu god Śiva is a favorite in Vraja because he is believed to be a great devotee of Kṛṣṇa. The following story told about him in Vraja demonstrates the requirement of a new body as a prerequisite for entering Kṛṣṇa's *līlā*.

One day Śiva, out of his immense love for Kṛṣṇa, desired to participate in the *rāsa-līlā,* the amorous circle dance that takes place in the groves of Vṛndāvana under the autumn moon. He was, however, unsuccessful in his attempts to enter Vṛndāvana.[90] He therefore consulted Yoga-māyā, the goddess in charge of creating the *līlā* of Vraja, and asked her what he should do to be able to enter Vṛndāvana and participate in the *rāsa-līlā.* Yoga-māyā informed Śiva that he could not enter Vṛndāvana with the body of a male ascetic; to enter Vṛndāvana he would first have to obtain the body of a woman, similar in form to that of the *gopī*s. To acquire this body, she advised him to bathe in a lake called Mānasarovara. There is a pun intended here. *Mānasa* means "mental" or "spiritual," and *sarovara* is a lake or pond. The lake at the base of Mount Kailāsa, sacred to Śiva, is called Mānasasarovara. However, this is not the lake Yoga-māyā was referring to, unless it moved to Vraja to be with Śiva in a different form. About three miles across the Yamunā River from Vṛndāvana lies a small lake. This body of water is surrounded by large beautiful green trees, which give it the feeling of an oasis-like sanctuary. Huge cranes and herons wade about the shallows and cows seek refuge in the shade of the lush trees. The people of present-day Vraja believe that this is the very lake that Rādhā came to in order to be alone and pout when Kṛṣṇa had gone off with another woman. The lake is thus called Mānasarovara. (*Māna* is a sulking anger or indignation caused by jealousy.) And this is the lake in which Śiva was instructed to bathe in order to obtain the *gopī*-body, the key to entering the amorous *līlā* of Vṛndāvana.[91] Śiva followed the advice of

Yoga-māyā, emerged from the water as a *gopī,* and was then able to cross the Yamunā and dance in the *līlā.*[92]

The commonly held notion in Vraja, that only a woman can enter and participate in the Vṛndāvana-līlā is further supported by a story told of the famous sixteenth-century female devotee of Kṛṣṇa, Mīrā Bai. The story relates how Mīrā Bai came to Vṛndāvana to meet Sanātana Gosvāmin[93] and sent ahead a message expressing her request. Sanātana, however, felt that a private meeting with a woman violated his disciplinary vows and sent her a message explaining that for these reasons he could not meet with her. The clever Mīrā retorted that she was quite surprised at his message, for it was a revelation to her that there were any males in Vṛndāvana except for Kṛṣṇa. She had assumed that all others residing there were females (i.e., *bhakta*s identified as *gopī*s). As a result of this rejoinder Mīrā was granted her desired meeting with Sanātana. The point of the story is that all taking part in the Vṛndāvana-līlā are females, whether this status is acquired by birth or through the Rāgānugā Bhakti Sādhana.

Not just any female body, however, will guarantee access to Vṛndāvana. A short distance upstream from Mānasarovara is a small grove by the name of Bel-ban (Sanskrit, *Vela-vana,* a mango tree grove). Within this beautiful grove, populated with brilliant peacocks, there stands a temple to Lakṣmī, the queen of Viṣṇu in the majestic Vaikuṇṭha. Inside Vṛndāvana proper, there exists no temple for Lakṣmī. This is because Lakṣmī was unable to enter Vṛndāvana, due to the fact that she would not follow the simple ways of the *gopī*s or adopt the *parakīyā* (unmarried) sentiment.[94] The story connected with the temple in Bel-ban narrates how the *gopī*s arranged for Lakṣmī to wait for them in the grove of Bel-ban, promising to return for her later in the night and lead her to the side of the *rāsa-līlā* in Vṛndāvana. However, since Lakṣmī would not show the *gopī*s the proper respect, they never returned for her and she was unable to cross the Yamunā River and enter Vṛndāvana without their help.[95] The lesson intended is that one who is unable to adopt the path of *sādhana* which follows the ways of the *gopī*s cannot be part of Kṛṣṇa's *līlā* in Vṛndāvana. The identity of a *gopī* (*gopī-abhimāna*) is a necessary prerequisite for participation in the amorous world of Vṛndāvana.[96] This *gopī*-identity, or *gopī*-body, is an example of what the Gauḍīya Vaiṣṇavas call the *siddha-rūpa* or *siddha-deha.*

The term *siddha-rūpa*[97] first appears in Gauḍīya Vaiṣṇava literature in the *Bhaktirasāmṛtasindhu* of Rūpa Gosvāmin. Rūpa says that service or religious performance (*sevā*) is to be done in a manner that imitates the Vrajaloka with both the "practitioner's body" (*sādhaka-rūpa*) and the "perfected body" (*siddha-rūpa;* BRS 1.2.295). Rūpa writes no more about the nature of these two bodies, and it is difficult to know exactly what they refer to from his work alone. The history of the Rāgānugā Bhakti Sādhana can almost be studied from the interpretive commentaries on this important yet ambiguous verse alone. The *sādhaka-rūpa* seems to denote the body with which the practitioner comes to the religious practice. This interpretation is borne out by Jīva's gloss of the *sādhaka-rūpa* as "the body as it presently is" (*yathāsthita-deha*).[98]

The *siddha-rūpa,* or *siddha-deha,* is a much more mysterious body than the *sādhaka-rūpa.* The term *siddha-deha* had a previous history in the Nāth and Haṭha-yoga traditions, where it denoted the body which had been perfected through techniques of *yoga.*[99] The term referred to the ordinary body when it had attained a deathless state free from disintegration.[100] The Gauḍīya Vaiṣṇava tradition, however, uses the term in a technical manner, involving two ideas. The *siddha-deha* or *siddha-rūpa* is first a meditative body. It is sometimes called the *bhāvanāmaya-rūpa*—a body developed and used in meditation. In his commentary on the *Bhaktirasāmṛtasindhu,* Jīva glosses *siddha-rūpa* as "an inner meditative body suitable for one's desired performative acts toward him (i.e., Kṛṣṇa)" (*antaś-cintitābhiṣṭa-tat-sevopayogi-deha;* BRS 1.2.295). The *siddha-rūpa,* then, is the body of the role that the practitioner assumes in the *līlā* of his meditation.[101] It is the meditative body the *bhakta* conceives of inhabiting and acting in while participating in the cosmic drama of Vraja.

But the *siddha-rūpa* is more than a meditative body for the Gauḍīya Vaiṣṇavas. It is the eternal form one will inhabit in the eternal *līlā.* It is sometimes called the *siddha-svarūpa*—one's essential nature. In his *Prīti Sandarbha,* Jīva Gosvāmin has described the form (*rūpa*) or body (*deha*) given to the liberated souls who have attained Bhagavān Kṛṣṇa. In so doing, he also gives the scriptural authority for the Gauḍīya Vaiṣṇava understanding of the *siddha-rūpa.* He writes:

> The Lord makes a body for the liberated one that is identical to those eternal bodies of the people of Vaikuṇṭha (Vraja), which consist of a particle of light from the Lord of Vaikuṇṭha (Vraja).[102]

That is, after the *bhakta* attains liberation, the Lord will grant him or her an eternal body with which to reside in the eternal world. Once again we see that it is an identity/body that connects one with a particular world.

The scriptural authority for this statement comes from the *Bhāgavata Purāṇa.* Jīva cites two verses:

All the people living there are endowed with a form of Vaikuṇṭha (*vaikuṇṭha-mūrti*).

When I was transformed into a pure divine body (*bhāgavatī-tanu*), my earthly body in which past karmic effects had been extinguished fell away.[103]

Jīva cites also Śrīdhara's commentary on the latter verse. This authoritative commentary indicates that the "companion body" (*pārṣada-tanu*) of those acting in Kṛṣṇa's *līlā* is pure, eternal, and devoid of *karma*.

Hence, the *siddha-rūpa* is the eternal spiritual body one is to inhabit at the end of *sādhana* when perfection (*siddha*) has been attained. It is variously called the *siddha-rūpa, siddha-deha, bhāgavatī-tanu, vaikuṇṭha-mūrti,* and *mukta-puruṣa.* Previously, I have indicated that it is also the body the practitioners conceive of themselves as occupying during the *sādhana*—that is, during the meditative practices, worship, etc. But these are not two distinct bodies; they are one and the same. The body that one meditates in will be the very body that one resides in eternally after death. This notion is apparently motivated by beliefs such as those reflected in this verse from the *Bhagavad-gītā:*

Whatever state one meditates on, one goes to as one gives up the body at death, O Son of Kuntī. One is then ever formed in that state (8.6).

In the *Rāgavartmacandrikā,* Viśvanātha Cakravartin explicitly states that if the practitioner's *sādhana* is successful, at the time of death he will be born again in the form of his *siddha-rūpa* as a *gopī* in Vraja. This form, he says, will consist of consciousness and bliss (*cid-ānanda-mayī gopikā-tanu*).[104]

Many Gauḍīya Vaiṣṇavas insist that the *siddha-rūpa* is indicated in the *Padma Purāṇa.* A passage from this Purāṇa briefly describes the form that the *siddha-rūpa* will take for most Gauḍīya Vaiṣṇavas who cherish the *gopī*s as the highest models of *bhakti.* It reads:

One should conceive of oneself there among them (i.e., in Vṛndāvana among the *gopī*s) as a charming adolescent woman of a delightful appearance, endowed with beauty and youth, skilled in various arts, and suitable for the enjoyment of Kṛṣṇa.[105]

The form described in this verse is the *gopī*-body in which the practitioners in pursuit of the amorous emotional role are to conceive of themselves.

One of the earliest detailed descriptions of the *siddha-rūpa* by a known Gauḍīya Vaiṣṇava is found in Gopālaguru Gosvāmin's *Gaura-*

Govindārcana-smaraṇa-paddhati.[106] I translate his description of the *siddha-deha* in full.[107] The *siddha-deha* is:

> An adolescent female cowherder who is adorned with all kinds of ornaments and two large pointed breasts, and possesses the sixty-four qualities of beauty.
>
> She has a secret love for Govinda, is confused by the joy of that love, and is equipped with aesthetically pleasing speech and a divine form.
>
> She bears the well-known *bhāvollāsa,* which is the *rasa* of the female friend, and loves the two (Rādhā and Kṛṣṇa) in her heart night and day.
>
> She is complete with all decorations, adorned with a variety of emotions, born by the grace of the *guru*, follows the dear form (i.e., the *siddha-deha*) of the *guru,* and is situated in a Gāndharva-like group, such as Lalitā's.
>
> She follows Rūpa Mañjarī, dwells in the village of Yāvaṭa, and imitates the Amorous Bhakti of the model *gopī*s (*kāma-rūpā*). This is the form which is to be visualized.
>
> She consists of consciousness, joy, and *rasa,* has a splendor equal to the moon, is dressed in blue clothes, and wears various ornamental decorations.
>
> She is situated by the side of Rādhā-Kṛṣṇa, is fresh and young, is a *mañjarī* of a certain family, with a mother whose name is given by the *guru,* a father who is third in the family, and a husband who gives the family its name.
>
> She is soft and able, has a residence in Yāvaṭa, and is richly endowed with ornaments and clothing for Śrī Rādhā.
>
> And the eleven well-known and exceedingly charming aspects of this *siddha-deha* are defined in the following way:
>
> Name, beauty, age, dress, relationship, group, command, service, specialty, guardian *sakhī,* and residence.

This last verse is especially important, for it mentions eleven aspects that provide a structural framework for the *siddha-deha.* These eleven descriptive aspects of the *siddha-deha* or *siddha-rūpa* are what the *guru* reveals to the practitioner at the time of an esoteric initiation, thereby defining in great detail the specific character the practitioner is to assume in the cosmic drama.[108] The *siddha-rūpa* of Rūpa Gosvāmin himself is often described and can be presented as an example to give a better picture of the type of role the Rāgānugā practitioner typically pursues.[109]

The *Siddha-rūpa* of Rūpa Gosvāmin

Name:	Rūpa Mañjarī
Age:	thirteen years and six months
Complexion:	dark yellow
Color and pattern of sari:	eyes of peacock's tail feathers
Name of bower:	Candra (moon)
Service:	serving betel
Name of father:	Ratna
Place of father's residence:	Vṛṣabhānupura
Name of husband:	Durmada
Husband's residence:	Yāvaṭa

The Gauḍīya Vaiṣṇava understanding of the *siddha-rūpa* has remained fairly constant since its form was worked out in detail around the beginning of the seventeenth century.[110] Interviews with Rāgānugā practitioners in present-day Vraja and readings of modern Bengali works on the subject reveal a continuation of the idea that the *siddha-rūpa* is a supernatural (*aprākṛta*) body that houses the soul in the eternal world. It consists of a particle of light (*jyotir-aṃśa*) that comes from Bhagavān Kṛṣṇa and is comprised of consciousness and bliss (*cid-ānanda-mayī*). One's *siddha-rūpa* will be of a form appropriate to one's given emotional role (*bhāva*) in the cosmic drama. For example, the *siddha-rūpa* of a practitioner who is assuming the emotional role of a friend of Kṛṣṇa will take the form of an adolescent cowboy (*gopa*) and will be dressed with the clothing and ornaments suitable to that form.[111] The performance of this type of character involves such actions as arranging trysts for Kṛṣṇa and wrestling with him for his enjoyment. The most common form of the *siddha-rūpa,* however, is the *gopī*-body already described, for the amorous role is held to be supreme.

Stories of perfected practitioners (*siddha*s) and their adventures in their *siddha-rūpa*s are very popular in Vraja today; these make up much of the important lore which is passed around the various feasts and festivals by Rāgānugā practitioners. These stories provide further understanding of the nature of the *siddha-rūpa* and how it functions in the Rāgānugā Bhakti Sādhana. They are of particular interest when they relate how the *siddha-rūpa* affects the external life of the practitioner, for they demonstrate the potential influence on overall behavior of internal meditative experiences in the *siddha-rūpa.*

External manifestations of inner meditative experiences in the *siddha-*

rūpa are taken very seriously by the tradition. In fact, one of my informants in Vṛndāvana defined the perfected one, the *siddha,* as one who manifests something from the eternal *līlā* on the physical body. As an example, he related how one *bābā* (the renunciant of Gauḍīya Vaiṣṇavism) whose *siddha-rūpa* had the form of a dear friend (*priya-sakha*) was once playing Holi[112] with Kṛṣṇa. During the sport, Kṛṣṇa sprayed him with some colors. These colors were later visible to others on his physical body. This, my informant insisted, was proof that he was a *siddha.* Other examples I heard while in Vraja involved *siddha*s producing various objects from the world of the *līlā* that their *siddha-rūpa* had obtained. From time to time, I was told, a *bābā* whose *siddha-rūpa* was female was known to have grown breasts. These things, I continually was assured, were "clear facts." These stories add support to the understanding that the *siddha-rūpa* is the cornerstrone of participation in the world of the Vraja-līlā.

Many such stories are told in Vraja concerning one Siddha Kṛṣṇadāsa Bābā, who resided near Mount Govardhana in the early eighteenth century. Here are two of them. Kṛṣṇadāsa Bābā's *siddha-rūpa* was a young *gopī* in the service of Rādhā. One day he/she was placing bangles on Rādhā's arm, preparing Rādhā to meet with Kṛṣṇa. He became so absorbed in the *līlā* that to those around him it appeared that he had become unconscious for a period of about three hours. When he "awoke" and was asked what had happened, he remarked that from his/her perspective in *līlā*-time only moments had passed.[113]

The second incident is even more dramatic. Kṛṣṇadāsa Bābā was once again absorbed in his meditations as a female servant-friend of Rādhā, and noticed that Rādhā had dropped her ring into a pond. In an act of service to Rādhā he/she dove in after the ring. Those on the outside world witnessed Kṛṣṇadāsa Bābā dive into a nearby pond. He did not emerge for seven days. When he finally did come out of the water, he related that he/she had only momentarily been under water to recover the ring. Evidently *līlā*-time is not the same as ordinary time; a few moments in the *līlā* appeared to be seven days to those outside of it.

A related story is told of Raghunātha Dāsa Gosvāmin:

> One day Raghunātha was suffering from indigestion when one of his followers Viṭṭhalanātha called two physicians for treatment. After feeling the pulse they diagnosed the ailment to be due to eating of boiled rice with milk. Viṭṭhala laughed at this expression of opinion as to the cause of the disease, and threw doubts on their ability to examine the pulse correctly, saying that Gosvāmījī had not taken food at all. At this, Raghunātha intervened and stated that the physicians were correct, for the previous day, he had in

meditation (through his *siddha-rūpa*) offered boiled rice with milk to Rādhā Kṛṣṇa and ate the *prasāda* (the offered food) mentally.[114]

These stories demonstrate the Gauḍīya Vaiṣṇava belief that inner meditative experiences can have a tremendous effect on the shape of one's life. Moreover, they indicate that meditative, or we might even say imaginative (the word is *bhāvanā-maya*), experiences are real, perhaps more real than the "real" itself. The assumption that imaginative meditative experience does not create illusion, but rather illuminates reality, is quite common throughout South Asia.[115] In his commentary on the *Bhaktirasāmṛtasindhu* (1.2.182), Jīva illustrates the powerful effectivity of meditative experience. There he quotes a passage from the *Brahmavaivarta Purāṇa* which tells of a *brahman* who, in meditation, mentally prepared a sweet dish of milk and rice in a golden vessel to offer to Viṣṇu. Before he offered the dish to Viṣṇu he wanted to determine how hot the mixture was, and so dipped two fingers into the vessel. He immediately felt the pain of burnt fingers. When he came out of his meditation, he noticed that two of his physical fingers were burnt. A later writer, commenting on these verses, assumes that "the fact that the injury of the burnt finger became manifest externally demonstrates the reality of imaginative or meditative (*bhāvanā-maya*) experience."[116] Experiences of this type, involving the *siddha-rūpa,* are considered to be of a superior nature and are thus highly valued and accordingly pursued.

This discussion of the *siddha-rūpa* should be sufficient to show that the *siddha-rūpa* is the true religio-aesthetic center (*āśraya*) within and the true identity (*abhimāna*) that the *bhakta* is seeking. The *bhāva* is the general emotional role, exemplified by the original Vrajaloka, followed by the practitioner of the Rāgānugā Bhakti Sādhana, whereas the *siddha-rūpa* defines the particular and detailed identity an individual *bhakta* assumes in the *sādhana.* Each *bhakta* is ultimately a specific character in the cosmic drama of Vraja. The original *gopī*s, for example, present a general model for religious behavior, but the individual practitioner, while following that general model, assumes the part of a particular *gopī* with a particular name, complexion, dress, special talent, residence, and so forth. (A discussion of how one acquires a particular *siddha-rūpa* appears in Chapter 7.) Moreover, it is solely by means of this particular identity, which is patterned after one of the original Vrajaloka, that the practitioner attains the world of Vraja and comes to know and enjoy the ultimate *viṣaya,* Kṛṣṇa. The *siddha-rūpa,* then, is the "person in the part," cultivated by the Rāgānugā practitioner as a dramatic vehicle into the ultimate religious world.

6

Conflict in Interpretation: Debate and Development within the Sādhana

Hari! O Hari! When shall I attain such a state?
When shall I, abandoning this male body,
assume the body of a female
and apply sandalwood paste to the bodies of
the divine couple?

NAROTTAMA DĀSA ṬHĀKURA[1]

Every religious tradition that utilizes the imitation of divine models as a religious guide sooner or later faces a certain set of problems: What does it really mean to imitate the divine model? Is the imitation to be literal or symbolic, external or internal? If literal and external, what physical acts are appropriate? Is the divine model to be appropriated to one's own socio-historical milieu, or to be imitated even when incongruent with that milieu?

These are questions crucial to many religious traditions, and a variety of solutions have been proposed. The Theravāda monks literally and physically imitate the Buddha of the Pali Canon as they enter a monastery in Cambodia. Their ordination involves a physical enactment of a sacred ritual drama patterned after the Great Retirement of Buddha.[2] On the other hand, the Shingon Buddhist monks of Japan imitate Dainichi more internally and symbolically, using *mudrā*s, *mantra*s, and *maṇḍala*s.[3] The Cistercian monks, under Bernard of Clairvaux, summarily understood the ideal Christ-role as a life of humility, and interpreted the imitation of Christ as chastity, poverty, and obedience to an abbot.[4] The Penitentes of New Mexico, however, sought to imitate the Passion of Christ in literal detail, physically reenacting the events surrounding the crucifixion.[5] The Sioux of North America staged communal performances which were designed to be literal reproductions of exemplary visions. Black Elk, for example, employed tribal dancers, horses, props, paint, and elaborate costumes to create a visionary world in great detail, and assumed the role of the sacred Red Man in an attempt to embody this paradigmatic individual who restored the Sioux nation in his vision.[6]

Further examples could be added, but these should suffice to suggest some possible interpretive options for imitative activity. Questions of interpretation—external or internal, literal or symbolic—form a creative tension within the history of religions. Moreover, it is frequently the case that within pluralistic societies religious models differ from the accepted social roles of the society. (Some groups—"non-differentiated"—equate the social and religious models, while others—"differentiated"—define the two quite separately; regardless, all societies place limits on the distance one is allowed to deviate from the social models of conduct.) When this occurs, problems often arise. The greater the incongruity between a given social identity and a divine model, the greater the intensity of these problems.

The Gauḍīya Vaiṣṇava requirement that even men must somehow emulate the *gopīs*, female divine models, to attain the highest state is one of the most radical imitative demands of any religious tradition. But it is frequently the extremes which stand out in sharp focus and yield insights into the more common patterns of behavior which elude awareness. Throughout history, a variety of solutions have been created to aid in the religious quest that employs paradigmatic individuals. I will now show how the Gauḍīya Vaiṣṇava tradition struggled with the particular interpretive problems that developed because of the great incongruity that exists between the aspiring imitators and the divine models, the Vrajaloka.

The Issue

Gauḍīya Vaiṣṇavas agree that the ultimate religious goal is to be transformed into a character in Vraja, usually a *gopī,* by imitating the original characters of Vraja, the Vrajaloka, or rather, since one's true identity is a *gopī,* to overcome—by imitating the original *gopīs*—the ignorance which keeps one from realizing that fact. However, Gauḍīya Vaiṣṇavas begin to disagree greatly over what exactly the imitation of the Vrajaloka means. From what has been said thus far, the reader might assume that the practitioners of the Rāgānugā Bhakti Sādhana in Vraja physically imitate the *gopīs* in all ways—dress, external behavior, and so forth. Though this does indeed occur, it is rare. Instead, we commonly find the full-time practitioners dressed as renunciants: head shaven except for the sacred tuft (*śikhā*), their only apparel a small white cloth around the waist. These claim that they are imitating the Vrajaloka in all ways, even physically. We will examine the specific interpretation that enables them to make such a statement. Yet there are others who make the same claim and even dress as *gopīs*. This group, as I have said, is rare, but the fact that contemporary Gauḍīya Vaiṣṇavas are eager to voice warnings such as the following indicates that such interpretations are still a matter of concern today.

> It is very important to stress that the performance (*sevā*) with the inner meditative *siddha-deha,* complete with dress, ornaments, etc., is only to take place in the mind. This performance is not for the existing practitioner's body (*sādhaka-deha,* i.e., the physical body), nor is the dress of the *siddha-deha* to be taken on with the practitioner's body. If a male practitioner following the emotional role of a female lover dresses up his existing male body with the clothes and ornaments of a beautiful woman, or if a female practitioner following the emotional role of a male companion dresses up her female body with the clothes and ornaments of a man, that practitioner is greatly deluded and produces ill effects.[7]

One of the central instructions of the Rāgānuga Bhakti Sādhana states that: "The one desirous of attaining one of the emotional states of the Vrajaloka should do performative acts of service (*sevā*) in a manner which imitates the Vrajaloka with both the "perfected body" (*siddha-rūpa*) *and* the "practitioner's body" (*sādhaka-rupa;* BRS 1.2.295). Clearly some type of imitative action was intended even with the physical body. Rūpa, however, says no more about the prescriptions and limitations of this imitative action. The laconic nature of Rūpa's instructions spawned conflicting interpretations, which primarily focused on just what was meant by the physical imitation of the divine models, the Vrajaloka.

The conflicts in interpretation did not extend to imitative action with the *siddha-rūpa*. Everyone seems to have accepted Jīva's gloss of this term as the "inner meditative body" (*antaś-cintita . . . deha*) with which the practitioner imitates the way of the *gopīs* in every detail—dress, behavior, and so on—in the meditative activities that take place in a mental world. Many contemporary Gauḍīya Vaiṣṇavas insist that the imitation of the Vrajaloka is to be strictly mental. Yet again Rūpa's instructions clearly state that it is somehow also to be done with the *sādhaka-rūpa,* which Jīva glossed as the "body as it is" (*yathā-sthita-deha*)—that is, the ordinary physical body. The problems arose in considering what it means to follow the Vrajaloka with this body, the physical body.

This issue did not seem to be a particular problem for Jīva; he hardly comments on it. Clearly he thinks *bhakti* is more than a mental act. He states in his commentary on the *Bhaktirasāmṛtasindhu* (1.1.11) that *bhakti* is comprised of physical, vocal, and mental acts (*kāya-vāṅ-mānasīyas tattac-ceṣṭa-rūpaḥ*). We may get a further hint of his ideas on this issue by looking at his *Bhakti Sandarbha,* in which he says that the Rāgātmikā Bhakti of the Vrajaloka involves a great amount of listening (*śravaṇa*), praising (*kīrtana*), remembering (*smaraṇa*), service (*pāda-sevana*), worship (*vandana*), and complete surrender of the self (*ātma-nivedana*).[8] These are the commonly accepted physical, vocal, and mental acts of *bhakti* listed among the injunctions of Vaidhī Bhakti, and are easily carried out by a male body. Jīva remarks that these acts are part of the *bhakti* of the exemplary Vrajaloka. Performing them, then, would be an imitation of the Vrajaloka. Perhaps what Jīva had in mind was for the Rāgānugā practitioner to imitate the Vrajaloka outwardly by performing these particular acts with the physical body, the *sādhaka-rūpa,* and inwardly by taking on the dress, behavior, and other traits of the Vrajaloka with the *siddha-rūpa.* It is difficult to know exactly what Jīva means from the few words he wrote on the matter, but this does seem to be the way his star pupil Kṛṣṇadāsa Kavirāja interpreted the *sādhana.* Kṛṣṇadāsa writes:

> The *sādhana* is of two kinds: external and internal. The external is performing listening (*śravaṇa*), praising (*kīrtana*), and so forth, with the *sādhaka-deha.* The internal is meditatively performing service to Kṛṣṇa in Vraja night and day in the mind with one's own *siddha-deha.*[9]

Questions concerning the proper method of imitating the Vrajaloka with the *sādhaka-rūpa,* the physical body, were not firmly settled by either Rūpa or Jīva, and over the course of time a variety of interpretations arose and came into conflict. In particular, two strategies developed to deal with the incongruity of the female models and the male practitioners. (I mention in passing that this is the way the texts speak of the incongruity. However, it is possible that a woman in this world is a male in the other. In either case, the problems are the same.) The two strategies eventually came into conflict, with the result that the first strategy I will speak of was soundly condemned by the orthodox tradition. This first strategy followed the seemingly logical development of the Rāgānugā Bhakti Sādhana and encouraged its male adherents to actually transform the physical body to be congruent with the female models, whereas we will see that the second strategy developed an interpretation which involved two different sets of models for the two different bodies. The practitioners following the first strategy literally and physically imitate the *gopīs* by taking on the dress and behavior of a woman. They believe that since their true and essential identity is a *gopī,* they should dress and act the part.[10] Many of the early followers of this path must certainly have had their own strong rationale for so doing, but they have left no written records. What we do have, however, are the words of Rūpa Kavirāja, a seventeenth-century writer who is usually blamed for giving written rationalization for such literal imitative action with the physical body. It is to his works, then, that we must turn to understand more fully this strategy.

Rūpa Kavirāja and Imitation of the *Gopīs*

No recent study of Gauḍīya Vaiṣṇavism mentions Rūpa Kavirāja.[11] There is a reason for this; his works were condemned by a council held in Jaipur in 1727. The resulting judgment of the council effectively shielded both the future tradition and future scholars from Rūpa Kavirāja and his works. Yet we must conclude that if he were important enough that a council convened especially to condemn his works, he must have offered a significantly influential interpretation of the *sādhana.* Moreover, it is the opinions of Rūpa Kavirāja that Viśvanātha Cakravartin singles out to attack, before giving his own interpretive solution concerning the proper way to imitate the Vrajaloka with the *sādhaka-rūpa.* It seems that Rūpa Kavirāja's interpretive position was considered a substantial threat. Examination of this evidence reveals an important debate over the interpretation of the *sādhana,* occurring from the second half of the seventeenth century to the early part of the eighteenth century. One side of the debate is represented by Rūpa Kavirāja, the other by Viśvanātha Cakravartin, who is usually credited with the definitive solution.

Very little information on the life of Rūpa Kavirāja exists.[12] The great Gauḍīya scholar Haridāsa Dāsa claims that Rūpa Kavirāja was a disciple of Śrīnivāsa Ācārya, the student of Jīva Gosvāmin.[13] Another scholar mentions that Rūpa Kavirāja was a disciple of Hemalatā Thākurāṇī, the illustrious daughter of Śrīnivāsa Ācārya.[14] Curiously, Haridāsa Dāsa endorses the latter view in his translation and commentary on the *Bhaktirasāmṛtasindhu.*[15] No sources are given, but the weight of these scholars makes it seem likely that Rūpa Kavirāja was the disciple of Śrīnivāsa or Śrīnivāsa's daughter, and that he lived around the middle of the seventeenth century. He must have lived most of his life in Vṛndāvana, until he was expelled from the Gauḍīya Vaiṣṇava community there and moved to Assam.

Two Sanskrit works of Rūpa Kavirāja survive. The first, entitled the *Sārasaṅgraha,* has been edited and published from four remaining manuscripts; the second, entitled the *Rāgānugāvivṛtti,* exists in manuscript form in the Vrindaban Research Institute.[16] Our present concerns can best be pursued by examining the entire *Rāgānugāvivṛtti* and a section of the *Sārasaṅgraha* called "The Four Sādhanas."[17] These two expositions are commentaries on Rūpa Gosvāmin's verse instructing the Rāgānugā practitioner to imitate the Vrajaloka with both the *siddha-rūpa* and *sādhaka-rūpa.* Rūpa Kavirāja's interpretation of the *sādhana,* his possible rationale for dressing the physical body as a female *gopī,* is as follows.

There exist four different types of *bhakti-sādhana*s, which lead to four respective kinds of heavenly realms (*dhāma*s). The first type of *sādhana* consists of engaging in acts of Vaidhī Bhakti with both the *sādhaka-rūpa* and the *siddha-rūpa.* This *sādhana,* Rūpa Kavirāja informs us, leads to Goloka, the realm in which the Lord displays his majestic and awesome form (*aiśvarya-rūpa*). Those who attain this realm worship him with awe and reverence.

The second and third heavenly realms are Dvārakā and Mathurā, the cities Kṛṣṇa inhabits as a prince after leaving the simple life of Vraja. Dvārakā is attained by means of a type of *sādhana* that follows the injunctions of Vaidhī Bhakti with the *sādhaka-rūpa* and imitates the Vrajaloka with the *siddha-rūpa.* Mathurā is attained by means of a *sādhana* which is a mixture of Vaidhī and Rāgānugā with both bodies. (Rāgānugā is here understood to mean "imitation of the Vrajaloka.") Rūpa Kavirāja grounds his theory in the verse from the *Bhaktirasāmṛtasindhu* that states: "He who has great amorous desire (for Kṛṣṇa), but acts only by means of the path of injunctions (*vaidhi-mārga*), becomes a queen in the city" (BRS 1.2.203). The practitioner this verse refers to, Rūpa Kavirāja asserts, has mixed Rāgānugā and Vaidhī. True Rāgānugā necessarily involves desire (*lobha*) and imitation of the Vrajaloka (*vrajalokānusāra*) with *both* the *siddha-rūpa* and the *sādhaka-rupa.* Anything less than pure Rāgānugā with both bodies will lead the practitioner to either Mathurā or Dvārakā, where he will become a wife of Kṛṣṇa and will enter into a relationship marked with the distance of respect.

The highest goal, however, is to attain the state of a *gopī* in Vraja. Vraja is the place where Kṛṣṇa manifests his lovable and infinitely approachable form (*mādhurya-rūpa*), which makes possible the profoundest love. Throughout his works, Rūpa Kavirāja stresses that the realm of Vraja can be attained only by means of a *sādhana* that imitates the Vrajaloka with both the *siddha-rūpa* and the *sādhaka-rūpa.* "Therefore, it is established that Vraja is attained by Rāgānugā (imitation of the Vrajaloka) with the inner *siddha-rūpa* and the outer *sādhaka-rūpa.*"[18] To understand better what Rūpa Kavirāja means by this, we must examine his definition of the various physical and spiritual bodies. The real key to understanding Rūpa Kavirāja's theory is to realize that for him there are two kinds of physical bodies.

Rūpa Kavirāja defines the *siddha-rūpa* as "that body which imitates (*anukāri*) the people of Vraja and has attained association with Śrī Kṛṣṇa and his intimates."[19] That is, Rūpa Kavirāja agrees with Jīva and others that the *siddha-rūpa* is a body similar to the original characters of the Vraja-līlā. He further takes it to be a meditative body; he calls it the *bhāvanā-maya-rūpa* and the *antaś-cintita-deha.*

He next defines two types of physical bodies. The first he calls the *taṭastha-rūpa*[20], defined as "that body which is attached to the state of the body as it is—male, etc.—and is connected to the worship of Śrī Kṛṣṇa in his majestic form. It is not to imitate the Vrajaloka."[21] The *taṭastha-rūpa,* then, is simply the ordinary physical body as it presently exists; it is male or female depending on one's physical form. Note that one is not to imitate the Vrajaloka with this body. Elsewhere it is called the "body as it is" (*yathāsthita-deha*), the term Jīva offered as a synonym for the *sādhaka-rūpa.* Here Rūpa Kavirāja deviates from the later writers who insist that the *sādhaka-rūpa* and the "body as it is" (*yathāsthita-deha*) are one and the same, and who will condemn his notion that there is a distinction between the two.

Rūpa Kavirāja's definition of the *sādhaka-rūpa* is fascinating and innovative. In the *Rāgānugāvivṛtti,* he outlines the difference between the ordinary physical body and the *sādhaka-rūpa.* He explains that the ordinary body (*taṭastha-rūpa*) is "a body in which the meditative form which imitates the people of Vraja and associates with Śrī Kṛṣṇa and his intimates has not been attained, but the ordinary body resides in the mind"; whereas the *sādhaka-rūpa* is "a body in which the meditative form which imitates the people of Vraja and associates with Śrī Kṛṣṇa and his intimates has been attained in the mind of the ordinary body."[22] The *sādhaka-rūpa,* according to Rūpa Kavirāja, is the physical body with a meditative body identical to the Vrajaloka residing within it. It is the physical body after it has been transformed by means of an esoteric initiation, undertaken by advanced practitioners in which the *siddha-rūpa,* or *gopī*-identity, has been imposed (*āropaṇa*) upon it.[23] "The meditative form which imitates the Vrajaloka is imposed on the *sādhaka-rūpa.*"[24] The initiatory imposition ontologically changes the practitioner's body. Rūpa Kavirāja states this very clearly. "The *sādhaka-rūpa* is not a condition of the ordinary body (*taṭastha-rūpa*), but is another body altogether."[25] The initiation process is compared to an alchemical transformation. " 'As iron turns into gold with the introduction of mercury, so a man is born again with the introduction of initiation.' From these words (of the *Haribhaktivilāsa*) we know that another body exists after the Vaiṣṇava initiation."[26] Thus the initiated practitioner's body, the *sādhaka-rūpa,* is no longer the same as the ordinary body and therefore, Rūpa Kavirāja was to maintain, is no longer subject to the rules of the ordinary body. In fact, Rūpa Kavirāja was to contend that the initiated practitioner must not follow the rules of ordinary society with this body, but rather is to follow the Vrajaloka *in all ways* (*pratyekam*).

The imposition of the *siddha-rūpa,* the meditative *gopī* form which imitates and associates with the Vrajaloka, eventually produces in the *sādhaka-rūpa* what Rūpa Kavirāja calls the "inner state" (*antar-daśā*). "The inner state in the *sādhaka-rūpa* is caused by Rāgānugā in the *sādhaka-rūpa,* which in turn is caused by imposing onto the *sādhaka-rūpa* the meditative form which imitates the people of Vraja."[27] The goal envisioned here is a particular state in which the practitioner takes on the identity of a *gopī* in even his *sādhaka-rūpa.* Rāgānugā with the *sādhaka-rūpa* is best summed up by Rūpa Kavirāja with these words: "Imitating the Vrajaloka with the *sādhaka-rūpa* means ceasing both to think of oneself as a male and to think of Kṛṣṇa as majestic, while still in the ordinary body."[28]

Rūpa Kavirāja outlines four states[29] for the practitioner, which illustrate the progression toward the ideal inner state of the *sādhaka-rūpa:* (1) the outer state (*bāhya-daśā*), a condition in which the identity is associated with the external form—e.g., "I am a male *brahmin*"; (2) the half-inner half-outer state (*antar-bāhya-daśā*), a condition in which the identity maintains the dual association of the inner and outer form—e.g., "I am externally a male *brahmin* and internally a female *gopī*"; (3) the state just prior to the inner state (*pūrvāntar-daśā*), a condition which immediately precedes complete identity with the inner *gopī* form—e.g., "I am a *gopī* with some remaining association with this male *brahmin* form"; and (4) the supreme inner state (*parāntar-daśā*), a condition in which complete identity with the true form has been achieved—e.g., "I am a *gopī!*" The developmental direction formulated here is clearly the intentional shifting of identity in the physical body to the ultimate "body" defined as a *gopī*.

Rūpa Kavirāja further argues that at the highest level, the supreme inner state, the performance (*sevā*) with the *siddha-rūpa* and the performance with the *sādhaka-rūpa* merge and take the same form. To explain this phenomenon, he utilizes a fascinating comparison between a vīṇā (Indian lute) and a vīṇā player on the one hand, and the *siddha-rūpa* and the inner state of the *sādhaka-rūpa* on the other:

> The *siddha-rūpa* and the *sādhaka-rūpa* are similar to a vīṇā and a vīṇā player. Even though the two [vīṇā and vīṇā player] are distinct there is a oneness of their songs, because their essence is similar; just so, even though the two bodies are distinct their performances (*sevā*) are similar and even simultaneous. As the song produced on the vīṇā is situated in the mind of the vīṇā player; so the performance which occurs in the *siddha-rūpa* is situated in the *sādhaka-rūpa.* When separated there is no *rasa* in the music of the vīṇā and vīṇā player; likewise, when separated there is no Vraja-bhāva born in the performance [of the *siddha-rūpa* or the *sādhaka-rūpa*].[30]

The main ideas expressed by these verses are that the religious performance (*sevā*) must be done with both the *siddha-rūpa* and the *sādhaka-rūpa* to gain the realm of Vraja, and that at the highest level of the inner state, the performance with the two bodies is identical. As the musician and his instrument become one during the music, and—to extend the metaphor to the drama—as the actor and his part become one during the performance, so the *sādhaka-rūpa* and *siddha-rūpa* become one during the ritual performance of *sevā*. Elsewhere Rūpa Kavirāja says:

> Hence, whatever is done with the *siddha-rūpa* takes place in the *sādhaka-rūpa,* and whatever is done with the *sādhaka-rūpa* takes place in the *siddha-rūpa.* Therefore, one must imitate the Vrajaloka with the mind, voice, and body of both the *siddha-rūpa* and the *sādhaka-rūpa.*[31]

One can begin to see how such statements could be used to justify and even encourage an imitation of the *gopī*s in all ways with the physical body.

Rūpa Kavirāja insists again and again that the imitation of the Vrajaloka must be external as well as internal (*antar-bahir vrajalokānusāraḥ*), and also that the external and internal forms are the same, that is, that they both imitate the Vrajaloka. He does not accept the interpretation which insists that the practitioner imitate the Vrajaloka only in the mind and follow the injunctions of Vaidhī Bhakti with the body. This procedure, he informs us, would lead only to Dvārakā, not to the intended goal, Vraja. In fact, Rūpa Kavirāja goes as far as to insist that "in this *sādhana* which imitates the Vrajaloka, the practitioner is forbidden to perform acts of Vaidhī with the *siddha-rūpa,* and even more so with the *sādhaka-rūpa.*"[32] Instead, he urges the practitioner to imitate the Vrajaloka with the mind, body, and voice of *both* bodies.

Rūpa Kavirāja recognized that advanced imitative performance with the *sādhaka-rūpa* could be mistakenly perceived by the uninitiated. He writes: "Actions which are being done with the practitioner's body (*sādhaka-rūpa*) are often perceived by others as actions being done with the ordinary body (*yathāsthita-deha*); so too actions being done with the body that has left the ordinary system of socio-religious duties (i.e., the *sādhaka-rūpa*) are often perceived by others as actions being done with the body which is still connected to the system of ordinary socio-religious duties (i.e., the *yathāsthita-deha*)."[33] What Rūpa Kavirāja seems to be saying here is that the initiated practitioner's body still resembles the ordinary body to the uninitiated. The uninitiated, not understanding that the practitioner's body is now of a different ontological nature, mistakenly perceive the actions of the advanced practitioner (such as, perhaps, dressing and decorating this body as a female) and thus find fault with them.

Rūpa Kavirāja did warn, however, that one is not to imitate the Vrajaloka or *gopīs* with the uninitiated physical body. The ordinary body is still bound to the rules of ordinary society. Moreover, Rūpa Kavirāja had no illusions concerning the limitations of the physical body. He maintains that, while the *siddha-rūpa* is ultimately free, the practitioner's body never completely loses its connection with the ordinary form. "The difference between the inner state in the *sādhaka-rūpa* and the inner state in the *siddha-rūpa* is that the *sādhaka-rūpa* is always associated with the ordinary body (*yathāsthita-deha*), while the *siddha-rūpa* is not."[34] The *siddha-rūpa* and the *sādhaka-rūpa* do have this major difference, but the behavioral guide for both is the same—the Vrajaloka.

There is no direct evidence that suggests whether Rūpa Kavirāja did or did not participate in such acts as dressing as a *gopī,* but his theories did lend themselves to the rationale, and even impetus, for such acts. His theories state that an inner *gopī* form is to be imposed on the physical body, thereby creating a new body which is exempt from the ordinary socio-religious injunctions and is to imitate the Vrajaloka in all ways.

Many of the theories of Rūpa Kavirāja were rejected and condemned by the Jaipur council.[35] To study or teach either the *Sārasaṅgraha* or the *Rāgānugāvivṛtti* was declared a crime punishable by the Mahārāja of

Jaipur.[36] The specific ideas singled out for condemnation included his theory of the four *sādhana*s and the four heavenly realms. Those present at the council insisted that there is no difference between Goloka and Vṛndāvana; there is only one heavenly realm. But the point that caused Rūpa Kavirāja the most trouble was his contention that at the highest level of Rāgānugā, the performative acts of the *siddha-rūpa* and the *sādhaka-rūpa* are the same and are guided only by the behavior of the original Vrajaloka. The council condemned Rūpa Kavirāja's idea that the *sādhaka-rūpa* is not the ordinary body (*yathāsthita-deha*) and is therefore exempt from the ordinary socio-religious rules. Praxis, not doxy, usually constitutes the point of contention in Hinduism. The participants of the council further condemned Rūpa Kavirāja for saying that the constant and occasional duties (*nitya* and *naimittika karma*s) are forbidden in both Vaidhī and Rāgānugā (although, as I read him, he said only that these are not to be done at the higher level of Rāgānugā), and that Vaidhī Bhakti is forbidden in Rāgānugā. The judgment of the council was effective. Rūpa Kavirāja's interpretation of what Rūpa Gosvāmin meant by "imitating the Vrajaloka with the *sādhaka-rūpa*" was not to be accepted as authoritative by orthodox Gauḍīya Vaiṣṇavas; that honor was to go to Viśvanātha Cakravartin.

Viśvanātha Cakravartin's Solution

Much more is known about Viśvanātha Cakravartin than about Rūpa Kavirāja. Viśvanātha lived sometime between 1654 and 1754 and spent most of his life in Vṛndāvana, writing many important Sanskrit commentaries and original works on Gauḍīya Vaiṣṇavism.[37] He was recognized as one of the greatest authorities of his day and became the most influential interpreter of the works of Rūpa Gosvāmin.[38] His most important interpretive works on the Rāgānugā Bhakti Sādhana are his commentary on the *Bhaktirasāmṛtasindhu*[39] and a separate text entitled the *Rāgavartmacandrikā*.[40] I will discuss those parts of his works that pertain to our present concerns.

Viśvanātha confronts the opinions of Rūpa Kavirāja regarding imitative action with the *sādhaka-rūpa* most directly in his commentary on the *Bhaktirasāmṛtasindhu* (1.2.295). He writes:

> (Some say that) the word Vrajaloka refers only to Śrī Rādhā, Candrāvalī, etc. (i.e., the *gopī*s), who are situated in Vraja, and that all types of performative acts (*sevā*), including the physical, are to be done in a man-

> ner that imitates only them (i.e., imitates only the *gopīs*). Therefore, since taking refuge at the feet of a *guru,* the eleventh day fast, the worship of *śālagrāma* stones and the *tulasī* plant, etc., were not done by them, then those of us who follow them should not do these things either. This is the opinion of the contemporary Sauramya [followers of Rūpa Kavirāja living in Surama Kuñja in Vṛndāvana] which has been rejected.

Viśvanātha does not agree in the least with Rūpa Kavirāja's contention that at the level of Rāgānugā the practitioner is exempt from the injunctions of scripture. Rather, he insists in the *Rāgavartmacandrikā:* "The one who claims that Rāgānugā Bhakti is always and completely beyond the injunctions of scripture . . . continually has experienced, experiences, and will experience ruin and is to be censured."[41] For Viśvanātha, the only difference between Rāgānugā and Vaidhī Bhakti with the *sādhaka-rūpa* is the motivating force. The actions of both paths are the same; the distinguishing feature of Rāgānugā is the desire (*lobha*) for the emotional position of a *gopī,* or perhaps of some other character of the cosmic drama.

> In fact, performative acts (*sevana*) which follow the path of injunctions, but are motivated by desire (*lobha*), are called Rāgānugā; and performative acts which follow the path of injunctions, but are motivated by those injunctions, are called Vaidhī. But performative acts which are not related to injunctions lead to destruction according to the declaration of revealed and remembered scripture.[42]

Viśvanātha here addresses the issue regarding what form actions with the physical body are to take. It is important for him to clarify this point as a means of counteracting those (such as Rūpa Kavirāja) who claim that the only guide for the external actions of the Rāgānugā practitioner is the actions of the original Vrajaloka, meaning the *gopīs*. Viśvanātha wanted to stop this trend once and for all by asserting that Rāgānugā for the physical body differed from Vaidhī only in terms of its motivation, not its form. Acts of Vaidhī Bhakti motivated by an intense desire, and not merely law, are Rāgānugā for Viśvanātha. As we have just seen, this is not so for Rūpa Kavirāja. Vaidhī Bhakti, even when motivated by an intense desire, is not at all equivalent to Rāgānugā for Rūpa Kavirāja. He emphasized that the distinguishing feature of Rāgānugā was "the imitation of the Vrajaloka" (*vrajalokānusāra*). Rūpa Kavirāja had even gone so far as to maintain that, since Rāgānugā did not even involve Vaidhī, the actions of the two paths were entirely different, even for the *sādhaka-rūpa*.

The weight of the term *vrajalokānusāra* (imitation of the Vrajaloka) in Rūpa Gosvāmin's definition of Rāgānugā, however, was too heavy for Viśvanātha to ignore. The question of what Rūpa meant by the imitation of the Vrajaloka with the *sādhaka-rūpa* remained problematic. The solution Viśvanātha came up with was a clever one, which seemed to solve the problem in a manner that still remains acceptable. I have seen no modern discussion of the Rāgānugā Bhakti Sādhana, in Bengali or Hindi, that does not invoke his strategy as an inherent part of the *sādhana*. Here is his solution.

The most crucial question for Viśvanātha was: Who are the Vrajaloka that the practitioner is to imitate with the *sādhaka-rūpa?* Viśvanātha had before him Jīva Gosvāmin's commentary on Rūpa's definitional verse presented in the *Bhaktirasāmṛtasindhu,* which instructed the Rāgānugā practitioner to imitate the Vrajaloka with both the *siddha-rūpa* and the *sādhaka-rūpa* (BRS 1.2.295). Jīva comments:

> One who is desirous of that special love of a lover of Kṛṣṇa living in Vraja should imitate the lovers of Kṛṣṇa, the Vrajaloka, as well as those who are following them.

Following this commentary, Viśvanātha writes:

> The lovers of Kṛṣṇa are Śrī Rādhā, Lalitā, Viśākhā, Śrī Rūpa Mañjarī, and other *gopī*s; those following them are . . . Śrī Rūpa, Sanātana, and the other Vṛndāvana Gosvāmins. All of them are to be imitated.
>
> Mental performance with the *siddha-rūpa* is to be done in a manner which imitates Śrī Rādhā, Lalitā, Viśākhā, Śrī Rūpa Mañjarī, and other *gopī*s. But physical performance with the *sādhaka-rūpa* is to be done in a manner which imitates Śrī Rūpa, Sanātana, and the other Vṛndāvana Gosvāmins.

Here we have Viśvanātha's solution stated in its most succinct form. It is repeated in the *Rāgavartmacandrikā.*[43] Viśvanātha's strategy features two sets of models for the two different bodies. From his perspective, Rūpa Kavirāja errs in advising the imitation of one set of models with the two distinct bodies. Viśvanātha readily accepted the *gopī*s as the models for the inner meditative imitation with the *siddha-rūpa.* But according to him the different bodies require different models. Viśvanātha holds up the "mythical" models—the *gopī*s such as Rādhā, Lalitā, Viśākhā, and Rūpa Mañjarī—for the practitioner to emulate in the inner meditative performance with the *siddha-rūpa.* However, and this is where Viśvanātha's theory is innovative, he holds up the "historical" models—the Vṛndāvana Gosvāmins, such as Rūpa and Sanātana Gosvāmin—for the practitioner to emulate in the outer physical performance with the *sādhaka-rūpa.*

This solution may strike us as a bit odd, since it is doubtful that Rūpa Gosvāmin, the designer of the imitative practice of Rāgānugā, meant for the practitioner to imitate him—namely, Rūpa Gosvāmin—as one of the Vrajaloka. Regardless, this was Viśvanātha's solution. Inner imitation of the *gopī*s with the meditative *siddha-rūpa* caused few problems, but the external application of this concept to behavior with the *sādhaka-rūpa* caused many. Viśvanātha recognized that the term Vrajaloka simply meant "the people of Vraja." Since Rūpa, Sanātana, and the other Vṛndāvana Gosvāmins resided in Vraja, Viśvanātha identified them as the exemplary Vrajaloka to be followed with the *sādhaka-rūpa.* The incongruity between the male practitioner and the female models thus disappears in Viśvanātha's solution, as he explains that the male practitioner is to imitate the female *gopī*s only with his own *siddha-rūpa,* which also is a female *gopī,* and is to physically imitate Rūpa Gosvāmin, a model congruent with the male practitioner. His solution was neat and lasting.

Viśvanātha's solution is not quite as simple as it might first seem, for it assumes a complex hagiographical development that coalesced around the figures of the Vṛndāvana Gosvāmins. It is well known that Caitanya was considered to be an incarnation of Kṛṣṇa, and later a dual incarnation of Rādhā-Kṛṣṇa. What is less well known, however, is that every major figure associated with Caitanya was considered to be a particular "mythical" character from the Vraja-līlā also incarnated as one of the various "historical" figures during the time of Caitanya. Most important for our concerns, Rūpa Gosvāmin came to be considered as the incarnation of Rūpa Mañjarī, an important *gopī* who served Rādhā and Kṛṣṇa in the mythical Vṛndāvana under the *sakhī* Lalitā. In Viśvanātha's list of the "lovers of Kṛṣṇa" whom the practitioner is to imitate with the *siddha-rūpa,* the name Rūpa Mañjarī appears. Rūpa Mañjarī is the *siddha-rūpa* of Rūpa Gosvāmin, the theologican who settled in the north Indian town of Vṛndāvana. The physical form he manifested in history is his *sādhaka-rūpa,* manifest not in his case to attain a perfected state through *sādhana* (for he is considered to be an eternally perfected one [*nitya-siddha*] by Viśvanātha), but rather to provide a physical model for the human practitioner. Therefore, Viśvanātha indicates, the practitioner is to imitate Rūpa's *siddha-rūpa* (a *gopī*) with his own *siddha-rūpa,* and Rūpa's *sādhaka-rūpa* (the exemplary aesthetic-ascetic who established and followed various injunctions himself) with his own *sādhaka-rūpa.* The solution is an intriguing weave of myth and history.

Viśvanātha's solution assumed and was constructed upon an existing theory of the *mañjarī.* It would therefore be useful to examine the figure of the *mañjarī* and the *sādhana* which developed in conjunction with it. This examination is additionally important because today the Rāgānugā Bhakti Sādhana is virtually synonymous with Mañjarī Sādhana.

Mañjarī Sādhana

Mañjarī[44] is the term the Gauḍīya Vaiṣṇavas use to designate a very particular type of role that has developed for participation in the cosmic drama of the Vraja-līlā. It is an important role for the modern Gauḍīya Vaiṣṇava, since today the greatest number of Rāgānugā practitioners follow the role of the *mañjarī*.[45] A *mañjarī* is a special kind of *gopī*. More specifically, the *mañjarī* is a particular kind of female friend (*sakhī*) to Rādhā, who serves the needs of the divine couple during their love play. It was mentioned in Chapter 4 that Rādhā had five kinds of friends.[46] Two among these, the friends close as life (*prāṇa-sakhī*s) and the eternal friends (*nitya-sakhī*s), are characterized as having a greater love for Rādhā than for Kṛṣṇa. (Though, of course, any love for Rādhā is dependent on Kṛṣṇa.) It is this type of *sakhī* that is typically identified as a *mañjarī*.[47] The *mañjarī* is a charming young *gopī* complete with all the qualities of beauty. She is always of an adolescent age (*vayaḥ–sandhi* or *nava–kaiśora:* twelve to fifteen years of age) since this is considered the period when emotions are capable of reaching the greatest intensity.[48] The *mañjarī* is further defined as being subordinate to or under the apprenticeship of a main *sakhī* such as Lalitā. However, the *mañjarī* is held to be superior to a regular *sakhī,* at least by those who follow the Mañjarī Sādhana, for the following reason. At the time when Rādhā and Kṛṣṇa commence their lovemaking, the regular *sakhī*s are required to leave. The *mañjarī*s, on the other hand, are allowed to stay and serve the needs of the divine couple, thus giving them direct access to witness their lovemaking. The desired end is a religious voyeurism, said to produce infinite bliss (*ānanda* or *rasa*). The types of service the *mañjarī*s render for the divine couple include serving betel nut, fetching water, fanning them after the heat of love, combing and braiding their hair, decorating their bodies, massaging their limbs, and entertaining them with music and dance.[49] Sometimes Rādhā even sends a *mañjarī* to Kṛṣṇa for his pleasure.[50]

The following poem by Narottama Dāsa Ṭhākura provides a good illustration of the mood of Mañjarī Sādhana, the process of assuming the role of a *mañjarī*.

Hari! O Hari! When shall I attain such a state?
When shall I, abandoning this male body,
assume the body of a female
and apply sandalwood paste to the bodies of
the divine couple?

Having drawn up your top knot of hair,
when shall I bind and encircle it with fresh *guñja* seeds,
string various flowers and offer them to you as a necklace,
assist the *sakhīs* in dressing your body with yellow cloth,
and place betel-nut in your mouth?

When shall my eyes be filled with the sight of the two forms
which steal away the mind?
This is my heart's desire.
Glory be to Rūpa and Sanātana (Gosvāmin).
This female body is my treasure.
Humbly, Narottama Dāsa.

Mañjarī Sādhana is not mentioned in the *Bhaktirasāmṛtasindhu* or any other text written by Rūpa Gosvāmin. It seems probable that Rūpa had more definite ideas in mind concerning the Rāgānugā Bhakti Sādhana, but he did not express them in any text that survives. Instead, he created a skeletal system with a wide range of possible transformative roles for practitioners striving to participate in the Vraja-līlā. Over the course of time, this skeletal frame was fleshed out through application and experimentation. The result was a particularizing of the wide range of optional roles into the Mañjarī Sādhana. This *sādhana* seems to be the logical extension of the search for a role which allowed one intimate access to witness the love-play of Rādhā and Kṛṣṇa. Moreover, it developed a new type of role that could be multiplied indefinitely, thus creating unlimited positions in the Vraja-līlā for the numerous practitioners.

Mañjarī Sādhana was not alluded to directly by any of the six original Vṛndāvana Gosvāmins, and there is no definite proof that Mañjarī Sādhana in its complete form was initiated by any of these Gosvāmins. Yet Niradprasād Nāth, a Bengali scholar of Narottama Dāsa Ṭhākura, convincingly argues that this *sādhana* is hinted at in a very seminal form in the poetic prayers of Rūpa Gosvāmin's *Stavamālā* and Raghunātha Dāsa's *Stavāvali* (which request the chance to serve Rādhā), though it was not fully developed until the early to mid-seventeenth century, the time of Narottama, who wrote many texts on Mañjarī Sādhana.[51] After Narottama's time, it was firmly established and became the widely accepted form of the Rāgānugā Bhakti Sādhana. If this is true, and evidence seems to support it, Rāgānugā Bhakti Sādhana developed into the specific form of Mañjarī Sādhana in that creative period of Gauḍīya Vaiṣṇavism in Vraja which followed the *Bhaktirasāmṛtasindhu* in the mid-sixteenth century and lasted until the careers of Narottama Dāsa and Śrīnivāsa Ācārya were over in the mid-seventeenth century. This move toward a narrower and more structured interpretation of the *sādhana* was reinforced by Viśvanātha Cakravartin at the beginning of the eighteenth century. Since this time, few changes have occurred in the orthodox understanding of the Rāgānugā Bhakti Sādhana.

The first text that mentions the *mañjarī* form is the *Gaura-gaṇoddeśa-dīpikā,* a Sanskrit text written by Paramānanda Kavikarṇapūra, the son of an elderly disciple of Caitanya.[52] This text was written in 1576 and provides a long list of the forms of the *siddha-rūpa*s of the more important figures in the developing movement. Significantly, the *siddha-rūpa*s of the six Gosvāmins are mentioned as *mañjarī*s. Kavikarṇapūra informs us that:

Rūpa Gosvāmin	=	Rūpa Mañjarī
Sanātana Gosvāmin	=	Rati or Lavaṅga Mañjarī
Gopāla Bhaṭṭa	=	Anaṅga or Guṇa Mañjarī
Raghunātha Bhaṭṭa	=	Rāga Mañjarī
Raghunātha Dāsa	=	Rasa or Rati Mañjarī
Jīva Gosvāmin	=	Vilāsa Mañjarī[53]

Narottama's *guru* Lokanātha Gosvāmin and the Vṛndāvana Gosvāmin's star pupil Kṛṣṇadāsa Kavirāja are usually added to the list of the six main Gosvāmins/*mañjarī*s, bringing the list to eight, the number of the chief *sakhī*s under which the *mañjarī*s serve. Kavikarṇapūra reveals Lokanātha's *siddha-rūpa* to be Līlā Mañjarī, but his text was written before the career of Kṛṣṇadāsa Kavirāja. Fifty or sixty years after Kavikarṇapūra wrote the *Gaura-gaṇoddeśa-dīpikā,* Dhyānacandra Gosvāmin, a disciple of Gopālaguru Gosvāmin, wrote a work called the *Gaura-govindārcana-paddhati,* in which he revealed the *siddha-rūpa*s of the Vṛndāvana leaders of the Gauḍīya Vaiṣṇava movement in the format of a *yoga-pīṭhāmbuja.*[54] His list is complete with all eight main *mañjarī*s.

Rūpa Gosvāmin	=	Rūpa Mañjarī
Sanātana Gosvāmin	=	Lavaṅga Mañjarī
Gopāla Bhaṭṭa	=	Guṇa Mañjarī
Raghunātha Bhaṭṭa	=	Rasa Mañjarī
Raghunātha Dāsa	=	Rati Mañjarī

Lokanātha Gosvāmin	=	Mañjulālī Mañjarī
Jīva Gosvāmin	=	Vilāsa Mañjarī
Kṛṣṇadāsa Kavirāja	=	Kasturī Mañjarī

Shortly after this Sanskrit text was written, Narottama Dāsa wrote a Bengali work entitled the *Rāgamālā*.[55] In this text he also reveals the *siddha-rūpa*s of the Vṛndāvana Gosvāmins. His list generally agrees with Dhyānacandra's, although he supplies the additional name Anaṅga Mañjarī for Gopāla Bhaṭṭa and curiously substitutes the *siddha* name Ānanda Mañjarī for his own *guru* Lokanātha, while elsewhere he refers to him by his more common *siddha* name, Mañjulālī Mañjarī. With occasional and slight variations, this is the list one finds on the *yoga-pīṭha*s of the practitioners in Vraja today. The present form of Mañjarī Sadhana was firmly established by Narottama Dāsa. These lists demonstrate that either the Vṛndāvana Gosvāmins themselves, or those who followed closely after them, thought of their *siddha-rūpa*s as *mañjarī*s and thereby planted the seed for the Mañjarī Sādhana.

During the creative period that followed the writing of the *Bhaktirasāmṛtasindhu,* a transformation occurred in the religio-aesthetic system of Rāgānugā Bhakti Sādhana concomitant with the development of Mañjarī Sādhana. In Rūpa's system, Kṛṣṇa is the sole *viṣaya* (religio-aesthetic object) and Rādhā, though supreme among the *āśraya*s and often portrayed as an unattainable ideal, is still herself considered an *āśraya* (religio-aesthetic vessel).[56] We note, however, that the *viṣaya* of Mañjarī Sādhana is not Kṛṣṇa alone, but rather Kṛṣṇa and Rādhā.[57] In the poetic prayers of Narottama Dāsa, for example, we find numerous requests for the chance to serve not Kṛṣṇa alone, nor even Rādhā alone, but Rādhā–Kṛṣṇa together.[58] Thus we see an important innovation in the devotional theory, which no doubt played a great part in the development of the Mañjarī Sādhana: Rādhā is raised from the position of a model *bhakta* or *āśraya* to the position of an appropriate object of devotion, a *viṣaya.*

This shift in Rādhā's position had a great effect on the understanding of the appropriate *sthāyi-bhāva,* or dominant emotion, to be cultivated in the life of *bhakti.* In Rūpa's system outlined in the *Bhaktirasāmṛtasindhu,* the love for Kṛṣṇa (*kṛṣṇa-rati*) is featured as the dominant emotion (*sthāyi-bhāva*) which evolves into *bhakti-rasa.* In the later works, however, a significant change can again be noticed. Narottama and other practitioners of Mañjarī Sādhana, as we have seen, direct their devotional efforts toward both Kṛṣṇa and Rādhā, with a particular emphasis on Rādhā. What in effect emerges from their works is a new *sthāyi-bhāva.* This new interpretation is grounded in a verse curiously added by Rūpa toward the end of his discussion of the *sthāyi-bhāva* of *bhakti-rasa* (BRS 2.5.128). There Rūpa writes:

> The love for the friend which is less than or equal to the love for Kṛṣṇa is a secondary emotion (*sañcārī*). If the love for the friend is greater than the love for Kṛṣṇa, then it is called *bhāvollāsa.*[59]

Jīva's commentary makes it clear that the "friend" referred to is Rādhā:

> If that love for the friend—belonging to Rādhā's girlfriends (*sakhīs*) such as Lalitā, and the others, who are the vessels (*āśraya*) of the mutual love which takes a vessel, the special chosen *bhakta* Rādhā, as the object (*viṣaya*)—is the same as or less than the love which takes Kṛṣṇa as the object, then it is called a secondary emotion (*sañcārī-bhāva*) of the love for Kṛṣṇa (*kṛṣṇa-rati*), due to the fact that the love for Kṛṣṇa is the basis of it and nourishes it. But in the amorous (*madhura*) *rasa,* wherever the love for the friend (i.e., Rādhā) is greater than the love which takes Kṛṣṇa as the object, and nourishes and dominates that love, it is considered a special kind of secondary emotion called *bhāvollāsa.* Although after remembering it Rūpa placed it here, it should have been added to the end of the section on secondary emotions, since it belongs to the same category as those.

For Jīva, the girlfriends of Rādhā are *āśraya*s who take an *āśraya* as their *viṣaya;* that is, they are vessels who take a vessel as the object of their love. And their particular kind of love is called *bhāvollāsa.* Later writers crucially involved in the development of the Mañjarī Sādhana make much of the term *bhāvollāsa.* It came to be known as the dominant emotion of the *mañjarī,* the *mañjarī-bhāva.* Throughout the writings of Gopālaguru Gosvāmin, Dhyānacandra Gosvāmin, and Narottama Dāsa, the term *bhāvollāsa* is frequently used to name the dominant emotion to be cultivated by the practitioners of Mañjarī Sādhana. Gopālaguru, for example, speaks of the *mañjarī* as "she who possesses the *bhāvollāsa* which evolves into the *rasa* of the female friend (*saṅgīta-rasa*)."[60]

A tendency can be noted already in these words of Gopālaguru, which, as far as I am aware, was not explicitly spelled out until recent times. In Jīva's commentary, the *bhāvollāsa* is still called a secondary emotion (*sañcārī-bhāva*), although it is said to be a very special kind of secondary emotion (*sañcāritve 'pi vaiśiṣṭyāpekṣayā*). But the words of Gopālaguru suggest that this emotion evolves into a particular kind of *rasa,* which he calls the *rasa* of the female friend (*saṅgīta-rasa*). However, a secondary emotion (*sañcārī* or *vyabhicāri-bhāva*), by definition, cannot evolve into a *rasa;* only a dominant emotion (*sthāyi-bhāva*) can do so. Kuñjabihārī

Dāsa, a twentieth-century theoretician of the Mañjarī Sādhana who resided in Vraja, makes this explicit in his own discussion of the aesthetics of the Mañjarī Sādhana and recognizes that such a claim amounts to the creation of a new dominant emotion or *sthāyi-bhāva.*

> The emotion of the *mañjarī* is not a secondary (*sañcārī*) or incidental (*āgantuka*) emotion; it is always a *sthāyi-bhāva.* Therefore, although only that love which takes Kṛṣṇa as its object is called a *sthāyi-bhāva* in the scriptures on *bhakti-rasa,* and although the love which is greater for Rādhā is not defined as a *sthāyi-bhāva* (nor a *sañcārī-bhāva*), the love for Rādhā by the ordinary *sakhī*s in the *madhura-rasa* can be called a new *sañcārī-bhāva,* and that great love of the *mañjarī*s for Śrī Rādhā, which is yet dependent on Śrī Kṛṣṇa, can be called a new *sthāyi-bhāva.*[61]

The aesthetic issue Kuñjabihārī Dāsa is struggling with is an important one. If the love for Rādhā is secondary to the love for Kṛṣṇa, as it is for the ordinary *sakhī*s, then it can properly be called a secondary emotion (*sañcārī-bhāva*), as Rūpa suggests. But if the love for Rādhā dominates the love for Kṛṣṇa, as in the case of special *sakhī*s known later as *mañjarī*s, what sense does it make to refer to the love for Rādhā as secondary (*sañcārī*) and the love for Kṛṣṇa as dominant (*sthāyī*)? Jīva tries to get around this problem by designating this emotion, the *bhāvollāsa,* as a special kind of secondary emotion. Kuñjabihārī Dāsa, on the other hand, goes on to call the *bhāvollāsa,* in which the love for Rādhā dominates all other emotions and can evolve into a *rasa,* what it is in fact functionally—a new type of *sthāyi-bhāva.* Mañjarī Sādhana, then, is based on this new *sthāyi-bhāva,* which instead of being a love for Kṛṣṇa alone, Kṛṣṇa-rati, involves a deep love for Rādhā. Yet because any love for a lover of Kṛṣṇa is always still dependent on Kṛṣṇa, the new *sthāyi-bhāva* should properly be understood as Rādhā-Kṛṣṇa-rati (though it is almost always referred to by the technical name *bhāvollāsa*), since it takes the dual *viṣaya* Rādhā–Kṛṣṇa as its object. One might go so far as to claim that the Mañjarī Sādhana is associated with a new *rasa;* this *rasa* could be named, following Gopālguru, the *rasa* of the female friend (*saṅgīta-rasa*). Kuñjabihārī Dāsa, however, is content to include the *bhāvollāsa* as simply another type of *sthāyi-bhāva* of the amorous (*mādhurya*) *rasa.*

If Rādhā is promoted to the position of an object of devotion, a *viṣaya,* who is the vessel, the *āśraya,* of this new *sthāyi-bhāva?* The answer, of course, is the *mañjarī,* for it is the role of the *mañjarī* that those devoted to Rādhā strive to assume in their religious practices. And significantly the foremost exemplar of all *mañjarī*s is Rūpa Mañjarī, the *siddha-rūpa* of Rūpa Gosvāmin.[62] Thus we witness what is not so uncommon in Indian religions—the "deification" of the creator of a religious system. Through the developments of Mañjarī Sādhana, Rūpa Gosvāmin has been transformed from the creator of the Rāgānugā Bhakti Sādhana to one of its paradigmatic individuals, the Vrajaloka. Viśvanātha Cakravartin makes it clear that the practitioners of the Mañjarī Sādhana are to imitate Rūpa Mañjarī with their *siddha-rūpa* and Rūpa Gosvāmin with their *sādhaka-rūpa*.

When I discussed with the practitioners of Vraja today Rūpa's verse that instructs the Rāgānugā practitioner to imitate the Vrajaloka with the physical body, many were eager to inform me that the term Vrajaloka in this verse means Rūpa Gosvāmin. They themselves, I was assured, were following the model established by Rūpa. A number of contemporary works expounding on Rāgānugā include a copy of a painting of Rūpa surviving from the sixteenth century. Rūpa is pictured with a shaven head, wearing only the small white cloth of an ascetic around his waist. This is how one finds many of the Rāgānugā renunciants (*bābās*) dressed in Vraja today. In their own mind, these *bābās* are literally imitating the Vrajaloka (Rūpa Gosvāmin), and not only in dress (perhaps this is only an insignificant detail), but in all ways. Thus the concept of imitating the Vrajaloka becomes a religious socializing factor as well as an inner meditative guide. To understand more about the external behavior of many serious practitioners of the Rāgānugā Bhakti Sādhana today, one would do better to examine the sacred biographies of the exemplary Rūpa Gosvāmin rather than the *gopīs* of the *Bhāgavata Purāṇa*. Viśvanātha Cakravartin and the previous developers of the Mañjarī Sadhana have made Rūpa say, in the spirit of St. Paul ("Be imitators of me, as I am of Christ" [1 Cor. 11:1]): "Be imitators of me, as I am of the Vrajaloka." The chain grows *ad infinitum*. The accomplished practitioner (*siddha*) who is following Rūpa, who is following the Vrajaloka, also becomes a model worthy of imitation.[63]

7

The Ritual Process: Rāgānugā in Action

I placed myself in the position of a milkmaid. I say to Shri Krisna, "Come! come to me only."

GODBOLE IN E. M. FORSTER'S *A Passage to India*

The theoretical framework of the Rāgānugā Bhakti Sādhana, primarily as set forth by Rūpa Gosvāmin in the *Bhaktirasāmṛtasindhu,* has been outlined in Chapter 5. Chapter 6 demonstrated that the interpretation and practice of this *sādhana* underwent significant change over time. A core of ideas, however, remained continuous. In this chapter I present selected examples of the ritual process of the *sādhana* as it is performed in Vraja today, utilizing both textual references and field notes I collected in interviews with the practitioners of the Rāgānugā Bhakti Sādhana in Vṛndāvana and Rādhākuṇḍa.

Rādhākuṇḍa[1] is a beautiful pond situated about sixteen miles southwest of Vṛndāvana. Red sandstone steps surround the pond and lead down into its mossy green waters. Large shade trees bursting forth in full foliage cast bright reflections on the surface of the pond and provide relief from the hot sun above. On one side of Rādhākuṇḍa a small tunnel connects the waters of this pond with Śyāmakuṇḍa (Śyāma = Kṛṣṇa). The sacred waters flow continually back and forth through a tunnel that connects the ponds of Rādhā and Kṛṣṇa. Śyāmakuṇḍa, too, is bordered by brilliant shade trees. Clustered along the shores of both ponds, which are collectively referred to as Rādhākuṇḍa, are a number of faded red and white brick *bhajana kuṭīra*s, the meditation huts used by the *bābā*s who reside in this holy place during their religious practices. *Bābā* is the term the Gauḍīya Vaiṣṇavas use to designate the serious male renunciant who has given up everything to pursue the spiritual life. In the early morning hours, the *bābā*s can be seen coming to the pond to bathe and offer flowers to Rādhā-Kṛṣṇa, for this is identified as the very pond where Rādhā and Kṛṣṇa meet for their midday love-play.[2] In his *Upadeśāmṛta,* Rūpa Gosvāmin explains that Rādhākuṇḍa is the most sacred of all the sites in Vraja and that to dwell there is immensely auspicious.[3] Rādhākuṇḍa is also known as the most beneficial place in all of Vraja to practice the Rāgānugā Bhakti Sādhana. Raghunātha Dāsa Gosvāmin, one of the original Vṛndāvana Gosvāmins who was heavily involved in the development and practice of the Rāgānugā Bhakti Sādhana, passed most of his years in Vraja performing this *sādhana* at Rādhākuṇḍa. His *samādhi* (tomb) stands nestled among the meditation huts on the north side of its shores. Rādhākuṇḍa is then a controlled environment, an ideal stage on which the Rāgānugā Bhakti Sādhana can seriously be pursued; it is thus the place for Rāgānugā Bhakti Sādhana par excellence. For this reason many *bābā*s choose to make it their home.

Although many householders in Vraja are involved in the Rāgānugā Bhakti Sādhana, and some very seriously, the *bābā*s living at Rādhākuṇḍa and elsewhere in Vraja are typically engaged in the practice at a much deeper level. These comprise only a small percentage of the overall population of Vraja. The practices I describe pertain chiefly to them, since their lifestyle allows time to pursue the *sādhana* more vigorously. The technique of *līlā-smaraṇa* meditation, for example, takes years of training and practice to perfect. The *bābā*s and householders of Vraja do, however, interact frequently in the many festivals, feasts, and temple ceremonies. Householders interested in the Rāgānugā Bhakti Sādhana regularly consult the *bābā*s for spiritual advice. Moreover, the *bābā*s, who are striving to follow the paradigmatic Vrajaloka, themselves embody those exemplary roles and thus make the divine models of the mythical world vividly present to the "lay" community of householders.

Initiation

Rūpa's words on the *siddha-rūpa* are so brief that from them alone we have very little idea how the *siddha-rūpa* was to be acquired. It may be that Rūpa thought the *siddha-rūpa* was simply produced in the process of *bhakti.* There is reason to believe that he might have held that the Vaiṣṇava *mantra* alone was sufficient to generate the *siddha-rūpa.*[4] It is also possible that one might perceive one's *siddha-rūpa* in a vision or dream. Nevertheless, the tradition was soon to contend that the knowledge of the *siddha-rūpa* was something that was usually acquired in initiation (*dīkṣā*). The process of transforming identity from the body as it is (*yathāsthita-deha*) to the ultimately real body (*siddha-rūpa*), therefore, begins with initiation under a qualified *guru*.

There are three types of *gurus* in the Gauḍīya Vaiṣṇava Sampradāya: the *śravaṇa-guru,* the *śikṣā-guru,* and the *mantra* or *dīkṣā-guru.*[5] The transformative process commences with an encounter with a *śravaṇa-guru,* any individual from whom the future practitioner first hears (*śravaṇa*) about the world of ultimate meaning. If a person is so inclined, merely hearing the stories of the Vraja-līlā produces a desire for one of the emotional states (*bhāvas*) therein expressed. If this desire develops to the point where the listener is eager to enter and participate in this world of ultimate meaning, he then seeks initiation (*dīkṣa*) into that world and instruction (*śiksā*) in the methods of permanently attaining it. The *dīkṣā-guru* provides the first; the *śiksā-guru* provides the second. All three types of *gurus* may be embodied by the same person, or they may be separate persons. There can be several *śravaṇa-gurus*, from whom the practitioner hears about spiritual matters, and several *śiksạ-gurus*, who give instruction in spiritual techniques,[6] but only one *dīkṣā-guru.* The *dīkṣā-guru* is clearly the most important of the *gurus*; once a practitioner takes initiation from a *dīkṣā-guru,* he is forbidden ever to abandon that *guru.*[7] The unique importance of the *dīkṣā-guru* will emerge in the following discussion.

The *dīkṣā-guru* is the *guru* proper in the common understanding of the term, since he or she is the one who initiates the practitioner into the spiritual life. The initiation imparted by the *dīkṣā-guru* is comprised of three stages. I learned from the practitioners I interviewed in Vraja that all three stages may be performed at one time, but it is much more common for the *gurus* to perform each successive stage of the initiation when they feel a particular practitioner is ready. There are, however, no hard and fast rules; the *dīkṣā-guru* determines the procedure. The initiation today is much less elaborate than that outlined in the sixteenth-century *Haribhaktivilāsa.*[8] The central event of the current initiation is the secret transmission of the *mantra* and the revelation of the *siddha* forms of the practitioner and his chain of *gurus*, which is known as the *siddha-praṇālī* (lineage of the perfected ones).

The first stage of the initiation is a very general one. It involves the whispering of the well-known and public *mahā-mantra*[9] of the Gauḍīya Vaiṣṇava Sampradāya into the ear of the initiate. The initiate is to meditate with this mantra until he is ready for the second stage of the initiation, the imparting of the secret and private *dīkṣa-mantra.* An ontological transformation is understood to take place at the moment the *guru* transmits the *dīkṣā-mantra* to the disciple. The *Haribhaktivilāsa* (Dhṛta Vacana 2.7) explains that the bestowal of the *dīkṣā-mantra* causes the receiver to be born again (*dvi-jatam*). Many believe that acquisition of the *mantra* creates a new body,[10] which must be developed and nourished by recitation of the *mantra.*

Jīva Gosvāmin cites a verse from the *Haribhaktivilāsa* which states that initiation yields knowledge of divine matters (*divyaṃ jñānam*). He glosses the terms "*divyaṃ jñānam*" as the knowledge of the essential form of the Lord (*bhagavat-svarūpa-jñānam*), and the knowledge of one's own particular relationship with the Lord (*bhagavatā sambandha-viśeṣa-jñānam*).[11] Knowledge of these aspects comes by means of the *mantra.* The words of the contemporary Bengali Gauḍīya Vaiṣṇava scholar Rādhāgovinda Nāth are relevant:

> Birth is of two kinds: ordinary and ultimate. Birth from the father's sperm and the mother's egg is ordinary, while birth from the *mantra-dīkṣā* is ultimate. The result of an ordinary birth is an ordinary relationship, that is, a relationship with the elders of the family such as the father and the grandfather. The result of the ultimate birth is an ultimate relationship, that is, a relationship with the elder *guru*s of one's own line, such as the immediate *guru* and the supreme guru, and by means of their kindness, a relationship with the Lord that suits one's own nature.[12]

The practitioner develops a general knowledge of the possible relationships or identities while listening to the stories of Kṛṣṇa and his passionate companions from the *śravaṇa-guru.* But in the initiation proper, the practitioner gains knowledge of the particular relationship that he or she is to have with Kṛṣṇa and his companions in the eternal *līlā.* This particular relationship is specifically defined in terms of the identity of the *siddha-rūpa,* the practitioner's eternal part in the drama.

Thus we see that the *mantra* gives the practitioner a new birth, a new identity, even a new body. With the *mantra,* the practitioner comes to know more and more about the ultimate world and his place in it. Moreover, it is the *mantra* that nourishes the growth of this new body with its particular identity and relationship with Kṛṣṇa. It seems very likely, then, that the early Vṛndāvana Gosvạmins thought that knowledge of the *siddha-rūpa,* with its inherent relationship to Kṛṣṇa, was acquired solely through the *mantra.* If this were the case, then the third stage of the initiation, the *siddha-praṇālī-dīkṣā,* would be unnecessary. And indeed, the early Gosvāmins make no mention of it. The contention that the third step (which involves the revelation of a chain of *gurus*) was not originally part of the initiation also seems plausible considering the fact that the Vṛndāvana Gosvāmins lived during the time of Caitanya, when there was as yet no long chain of *guru*s in their movement. They considered themselves to be living in the immediate presence of revelation.

Today, however, knowledge of the *siddha-rūpa* is typically acquired through an esoteric initiation called the *siddha-praṇālī-dīkṣā.* This is the third stage of initiation, in which the *guru* reveals the initiate's particular *siddha-rūpa,* his own *siddha-rūpa,* and the *siddha-rūpa*s of all the *guru*s in his lineage back to the time of Caitanya. The *siddha-rūpa* is here revealed in great detail, usually in terms of the eleven or twelve aspects mentioned in Chapter 5.[13] The first text that speaks of the *siddha-rūpa* as being revealed by the *guru* (*guru-datta*) is the early seventeenth-century *Gaura-Govindārcana-smaraṇa-paddhati* of Gopālaguru Gosvāmin.[14] The details of this initiation seem to have been worked out in that creative period contemporaneous with the development of Mañjarī Sādhana. The most common view held by the practitioners of Rāgānugā Bhakti Sādhana in Vraja today is that the *siddha-rūpa* is either revealed or acquired in the initiatory process of the *siddha-praṇālī-dīkṣā.*

My interviews with these practitioners brought to light the existence of two theories concerning the *guru*'s part in revealing the particular form of the initiate's *siddha-rūpa.* One theory might best be named the "inherent theory," and the other the "assigned theory." The adherents of the "inherent theory" contend that the *guru,* by means of his meditative vision, has the ability to see which character in the eternal *līlā* the initiate really is, essentially and at all times, though prior to initiation the person is unaware of the existence of this inner body. At the time of initiation, the *guru* reveals to the initiate the detailed form of his *siddha-rūpa* as seen in the world of the *līlā* by perfected meditators. The initiate then proceeds to execute performative acts which help him to discover this true identity for himself.

One informant related the following incident as a way of illustrating that one's *siddha-rūpa* exists always and can be perceived by experienced meditators.

> When I was young I went to the hut of an old *bābā* by the name of Gaurāṇga Dāsa Bābājī who was at that time living in Vṛndāvana. I found this *bābā* deep in meditation and decided to wait patiently for him to return. After a little while, Bābā muttered for me to help him from his hut to a chair under a veranda. I gave him my arm and assisted him to the veranda. I then sat quietly at the *bābā*'s feet. Some time later, Bābā came out of his meditative world and returned to this one. He soon asked, "Who helped me here?" I replied that I had. "But," said the *bābā,* "I was

> helped by a beautiful *gopī*. I touched her arm. Her skin was so beautiful and she wore beautiful bangles." Soon the *bābā* returned fully enough to this world that it dawned on him what had happened. He then explained to me that he had seen my essential nature (*siddha-svarūpa*), though I was unaware of its existence at that time.

A similar experience, recorded in the Hindi biography of a well-known *bābā* named Rāmadāsa Bābājī who lived in Vraja in recent times, was pointed out to me also as a way of demonstrating the view that the *siddha-rūpa* is an inherent part of a person, even though that person may be unaware of its existence. The incident is said to have occurred when Rāmadāsa as a young man was accompanying an older accomplished *bābā* by the name of Vrajavālā.

> One afternoon a strange thing happened. Vrajavālā and Rāmadāsa Bābājī went to bathe at Parvana Sarovara.[15] They picked a large rose from a nearby garden. As he entered the pond to bathe, Vrajavālā gave the rose to Rāmadāsa Bābājī and said, "Keep this, don't let anyone take it." Rāmadāsa Bābājī took the flower and sat down on the bank and began to meditate. Suddenly, there came the singing voice of a young woman. He turned around, looked behind him, and saw an adolescent girl *(kiśorī)* of a divinely brilliant form, coming toward him emitting light from her limbs. With her was a mature woman.
>
> The adolescent girl approached him and said laughingly, "That flower is mine. I picked it." Having said that, she snatched the flower from his hand and ran away. In a few moments she had disappeared.
>
> Vrajavālā had been watching all this. Returning quickly from the pond, he asked, "Where is the flower?" Bābājī Mahārāja answered, "An adolescent girl took it.
>
> "Did you recognize her?" asked Vrajavālā.
>
> "No," Bābājī replied.
>
> Becoming somewhat thoughtful, Vrajavālā, who was experienced in meditation, said, "That was your *siddha-svarūpa*."[16]

Rāmadāsa and Vrajavālā were on their way to a temple to offer the flower to the images of Rādhā and Kṛṣṇa. But the higher way to make such an offering would be to offer the flower directly (*sākṣāt*) to Rādhā and Kṛṣṇa with the *siddha-rūpa*. Presumably, this is what Rāmadāsa was doing without realizing it. The adolescent girl was his own *siddha-rūpa* engaged in the task of securing the flower.

Both of these incidents were told to me to illustrate that the *siddha-rūpa* is an inherent part of a person, though one is unaware of it before initiation. In both cases the *siddha-rūpa*s were perceived and made known by masters whose vision was perfected through meditation. The contention of this view is that the *dīkṣā-guru,* who must necessarily be a perfected master (*siddha*), enters the world of the *līlā* by means of meditation, ascertains the true form of the initiate, and then reveals that form to the initiate at the time of the *siddha-praṇālī-dīkṣā.*

The second theory I found operative in Vraja, the "assigned theory," argues that at the time of the initiation the *guru* assigns the initiate a *siddha-rūpa* appropriate to his nature. One *bābā* at Rādhākuṇḍa described this aspect of the *siddha-praṇālī-dīkṣā* in this way:

> The *siddha-rūpa*s exist eternally in the world of the *līlā* like shiny new cars awaiting a driver. That is, they lie dormant waiting for a soul (*jīva*) to animate them. The *guru,* as a manifestation of Bhagavān, determines in his meditations which body best suits the nature of the initiate and assigns him that particular body.

According to this perspective, the *guru* does not discern who the initiate inherently is; instead, the *guru* assigns the initiate an eternal body with which to play out a particular part in the eternal *līlā.* This view is endorsed by the contemporary Bengali scholar and practitioner of the Mañjarī Sādhana, Kuñjabihārī Dāsa. He writes:

> There exists eternally in the world of the Lord eternal bodies which are suitable for the service of the Lord. All these bodies are portions of the light of the Lord; that is, each body corresponds to each portion of His light. Therefore, they are like the body of the Lord, supernatural and composed of consciousness. These eternal bodies are clearly illustrated in the beautiful and auspicious bodies of the people of Vaikuṇṭha. All these bodies are companion bodies (*pārṣada-deha*s). At the time a soul attains final liberation, it receives one of these bodies which is appropriate to its level of love according to the wishes of the Lord. This is how the companion body [*pārṣada-deha,* another name for the *siddha-rūpa*] is obtained. All of these companion bodies are eternal. They exist eternally before being united with the liberated soul, and will exist eternally after that union, but before they are united with the soul they remain in an inactive state.
>
> Each and every eternal soul is a servant of the Lord, and for each and every one of them there exists a body suitable for its service in the world of the Lord. If by the grace of *bhakti* a desire for the service of the Lord is produced, then by the kindness of the Lord this body is attained.
>
> In the Gauḍīya Vaiṣṇava Sampradāya this body is introduced in the *siddha-praṇālī* which is received from the *guru.* It is not a figment of the imagination; it is eternal, and it is real. The *guru-deva,* having been

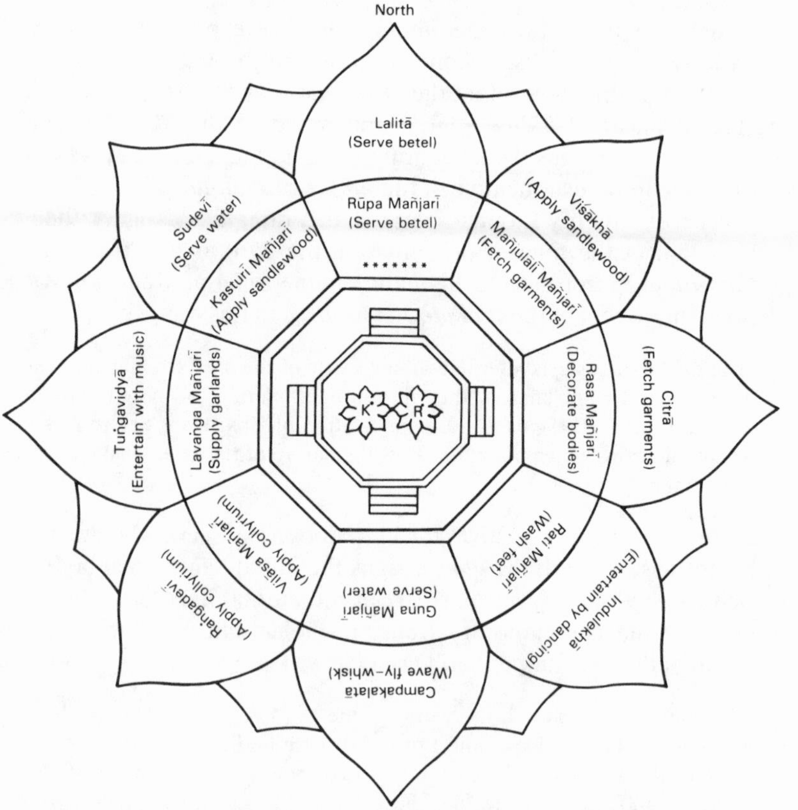

FIGURE 3. A *Yoga-pīṭhāmbuja* for Mañjarī Sādhana

informed in the state of meditation which of the eternal bodies existing in the world of the Lord the initiate is to be assigned by the Lord, reveals that body to the initiate as the initiate's *siddha-deha*.[17]

Regardless whether the *guru* reveals the initiate's existing essential nature or assigns the initiate a *siddha-rūpa* appropriate to his disposition, it is in the *siddha-praṇālī-dīkṣā* that the practitioner becomes aware of the particular part he is to play in the eternal drama. A diagram called a *yoga-pīṭhāmbuja*[18] is frequently used in the *siddha-praṇālī-dīkṣā* to locate the initiate in the drama with respect to the other characters, particularly his line of *guru*s. It is a device to facilitate the learning of the characters of the drama and one's relationship to them. A *yoga-pīṭhāmbuja* used in the Mañjarī Sādhana looks something like the diagram of Figure 3.[19] The diagram would typically be filled out with many of the colorful aspects of the *mañjarīs* mentioned previously.

The outer ring of characters of Figure 3 are the *sakhīs*, the inner ring are the *mañjarīs*. The asterisks indicate the initiate's line of *guru-mañjarīs*, the *siddha-rūpas* of the initiate's *dīkṣā-guru* lineage. In this particular case, the initiate would belong to the line of *mañjarīs* who serve under Rūpa Mañjarī (that is, in the line of Rūpa Gosvāmin), who in turn serves under the *sakhī* Lalitā.

The *dīkṣa-guru,* then, is extremely important to the transformative process, because he is the one who reveals the *mantra* and the *siddha-praṇālī* so essential to entering the world of the *līlā.* Some practitioners pass beyond the guidance of the *dīkṣā-guru* upon entering the *līlā* with the *siddha-rūpa* and put themselves directly under the guidance of the original *gopīs*, such as Lalitā and Viśākhā. For many practitioners, however, the involvement of the *dīkṣā-guru* is much more significant. The *dīkṣā-guru,* being a perfected master (*siddha*), is a *sakhī* or a *mañjarī* in the world of the *līlā.* Therefore, many practitioners conceive of themselves as a *sakhī* or *mañjarī* under the directorship of the perfected and exemplary *sakhī* or *mañjarī* form of their *guru* and execute their performance following the instructions and examples of the *guru*'s perfected form.[20] An eternal relationship with the *guru* is thus envisioned; since both the *guru* and the disciple possess an eternal body in the world of the *līlā,* the relationship begun in this world is extended into the other. In this way the *dīkṣā-guru, qua siddha,* is the *metteur en scene,* the dramatic director of the disciple's assigned performance.

Performance with the *Siddha-rūpa*

The Gauḍīya Vaiṣṇavas maintain that the salvific transformation of identity from the ordinary body to the ultimate identity of the *siddha-rūpa* is accomplished through acts of *sevā,* which I have translated as "religious performance." Rūpa Gosvāmin had stated that this performance is to be done with both the *siddha-rūpa* and the *sādhaka-rūpa* (BRS 1.2.295), and as previously mentioned, the most dominant interpretation prescribes two major types of religious performance for the practitioners: internal acts with the *siddha-rūpa* and external acts with the *sādhaka-rūpa.* Let us follow this division and begin our examination of Gauḍīya Vaiṣṇava performative techniques with the religious practices that take place in the inner mind.

Līlā-Smaraṇa Meditation

The inner or mental *sādhana* primarily involves a meditative technique known as *līlā-smaraṇa*. In one of the most important verses pertaining to the Rāgānugā Bhakti Sādhana, Rūpa instructs:

> (The Rāgānugā practitioner) should dwell continually in Vraja, absorbed in the stories (of the cosmic drama of Vraja), remembering (*smaraṇa*) Kṛṣṇa and his beloved intimates one is most attracted to (BRS 1.2.294).

The term *smaraṇa*, "remembering," "bearing in mind," or even "visualizing," is extremely prominent in Vaiṣṇava *bhakti sādhana*.[21] It is used throughout the *Bhagavad-gītā* as the technique of *bhakti* which enables the practitioner to reach Kṛṣṇa's abode.

> Whoever remembers (*smaran*) Me alone when leaving the body at the time of death attains my state. There is no doubt of this.
>
> Whatever state one remembers (*smaran*), one goes to as one gives up the body at death, O Son of Kuntī, for that state draws one to itself.
>
> Therefore, remember (*anusmara*) Me at all times and fight. With mind and intellect fixed on Me, you will certainly come to Me.
>
> He who meditates on the Supreme Person with a mind controlled by *yoga* and meditation, not wandering after anything else, goes to the Supreme Person, O Pārtha (8.5–8).

What seems to be assumed and expressed in these verses is the notion (so common throughout the history of Hindu thought) that one lives in one's mental projections. The world of mental images or imagination is taken much more seriously in India than it is typically in the West.[22] If one could somehow hold in mind (*smaraṇa*) a mental image harmonious with Ultimate Reality, one would live in or participate in (*bhakti*) that reality. One becomes what one "holds in mind." Therefore, the Vaiṣṇavas strive to meditate on, or remember, the Ultimate Reality conceived as the Divine Person, and in this way attempt to share in that reality.

The importance of the meditative technique of *smaraṇa* in Vaiṣṇava *bhakti* can further be observed in the definition of *bhakti* given by the famous eleventh-century Vaiṣṇava theologian Rāmānuja. In his commentary on the first of the *Brahma Sūtra*s, Rāmānuja engages in an inquiry regarding the saving knowledge of Brahman. He argues that it is knowledge realized in meditative experience rather than mere textual understanding that leads to liberation. Rāmānuja defines meditation (*upāsana*) as "a constant remembrance (*smṛti*), uninterrupted like the flow of oil."[23] He then goes on to specify that this meditation is a form of visualization. He writes:

> This remembrance (*smṛti*) takes the form of a vision (*darśana-rūpa*) and it possesses the character of immediate perception (*pratyakṣatā*). Remembrance is a form of direct perception (*sākṣātkāra-rūpa*).[24]

Thus when this remembrance becomes intense and perfected it results in a direct and vivid perception of Ultimate Reality, or the Supreme Person. This occurs when the practitioner has totally harmonized his mental projections with Ultimate Reality.

Rāmānuja declares that *smaraṇa,* this constant remembrance defined as a meditative technique of visualization, is what is meant by the term *bhakti.*[25] *Bhakti* then, for Rāmānuja, is a specific meditative technique of concentrating on an image of the deity visualized in the mind of the practitioner. He remarks in his commentary on the *Bhagavad-gītā* that this meditative technique of visualization is what Kṛṣṇa means when he says again and again: "Fix your mind on Me."[26]

Smaraṇa meditation, placed at the very center of *bhakti* by Rāmānuja, is a continuation of the yogic contemplative technique of concentration systematized in the *yoga-sūtra*s of Patañjali. The key difference between the two systems lies in the fact that in Rāmānuja's system the object of concentration is the Supreme Person Viṣṇu alone, while in Patañjali's system, the object of concentration might simply be a concept or sense-object of any sort.[27] The mental faculties in *bhakti* are focused exclusively on Viṣṇu, not with a view of making the object of meditation more and more abstract until an objectless state is obtained, as in classical yoga, but rather with a view of generating devotion toward Viṣṇu and eventually experiencing a direct vision of Viṣṇu.

Rūpa Gosvāmin includes remembrance (*smṛti*) among the list of sixty-four limbs of *sādhana* and defines it as "an association with the mind in whatever way" (BRS 1.2.175). Jīva Gosvāmin delineates five successive stages of *smaraṇa* for the Gauḍīya Vaiṣṇavas in his *Bhakti Sandarbha.*[28] He does so in terms that also strongly suggest a continuity with Patañjali's yogic techniques of contemplation. Jīva asserts that "remembrance" or "visualization" of Kṛṣṇa's qualities, companions, service, and *līlā*s is easily obtained by following the method of this sequential format. The first step is called simply *smaraṇa* and is defined as irregular reflection. The second is *dhāraṇa,* Patañjali's term for concentration, here defined by Jīva as a withdrawing of the mind from everything and fixing it on the object of meditation. The third step is *dhyāna,* Patañjali's term for meditation proper. Jīva defines it as a special meditation on the forms and other characteristics of Kṛṣṇa. Fourth is *drūvānusmṛti,* the term Rāmānuja used to define *bhakti,* as employed by Jīva it denotes a state of meditation that is similar to the previous *dhyāna* but uninterrupted, like a shower of ambrosia. The final step is *samādhi,* Patañjali's state of objectless consciousness, a term used by Jīva to designate the point in the practice of *smaraṇa* when the object of meditation itself appears. The goal of the *smaraṇa* meditation, then, for the Gauḍīya Vaiṣṇavas is to achieve a direct vision (*sākṣāt-darśana*) of Kṛṣṇa and his dramatic world. All theorists of Gauḍīya Vaiṣṇava practice, past and present, agree that *smaraṇa* is an essential and prominent feature of the Rāgānugā Bhakti Sādhana.[29]

Smaraṇa, specifically for the Gauḍīya Vaiṣṇavas, is a meditative technique of visualizing in the mind the *līlā* of Kṛṣṇa and his retinue of intimate companions in the enchanting land of Vraja. This technique is therefore usually referred to as *līlā-smaraṇa.*[30] The practice involves visualizing a particular dramatic scene of Vraja in great detail, establishing its setting (*deśa*), time (*kāla*), and characters (*patra*).[31] *Mantra*s are employed to assist the visualization. The practitioners memorize the descriptions of the various *līlā*s in an impressively elaborate manner, using maps and diagrams to locate the more important *līlā* activities. The mind is to be withdrawn from the ordinary world and completely concentrated on and absorbed in the *līlā* of Vraja. When this process is perfected, the cosmic drama appears directly before the eyes of the practitioner, granting visual access to the world of ultimate meaning.

Rūpa's instruction that the *līlā* is to be remembered continually is taken seriously by many practitioners. This instruction was taken to mean that the practitioner is to meditate on or visualize the *līlā* unceasingly throughout the entire day. The practitioner's day in Vraja is traditionally divided into eight periods, which constitute one day in Kṛṣṇa's *līlā.* The meditation which is structured after these eight time periods is called *aṣṭa-kālīya-līlā-smaraṇa.* Three verses of Gopālaguru Gosvāmin, often quoted by the practitioners of Rāgānugā, name the eight periods and indicate their length.

> Having first visualized Rādhā and Kṛṣṇa on a lotus, one should perform their service with the *siddha-deha* according to the eight time periods.
>
> The eight time periods are, in order: night's end, morning, forenoon, midday, afternoon, sunset, late evening, and night.
>
> Both midday and night are remembered for six *muhūrta*s, and night's end and the rest are known to be three *muhūrta*s.[32]

Aṣṭā-Kālīya-Līlā

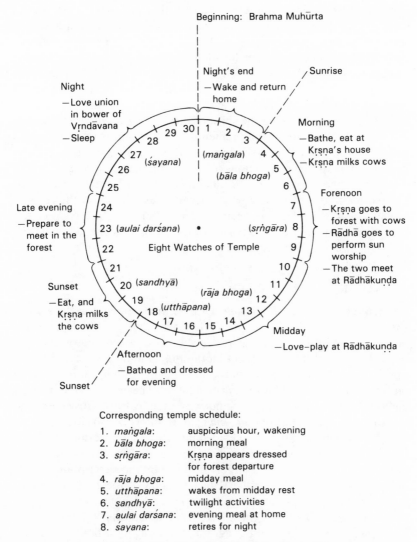

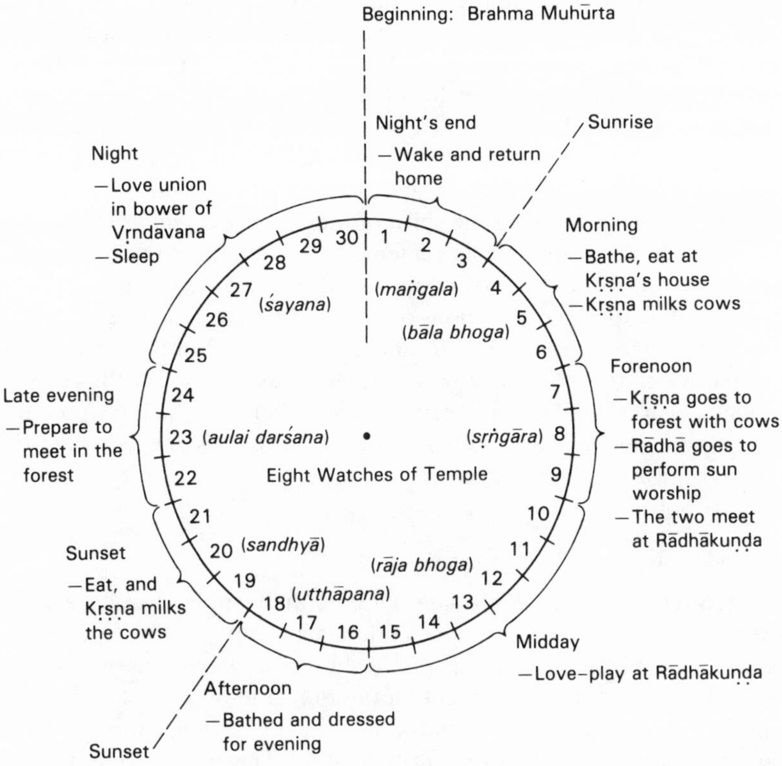

Corresponding temple schedule:

1. *maṅgala*:	auspicious hour, wakening
2. *bāla bhoga*:	morning meal
3. *sṛṅgāra*:	Kṛṣṇa appears dressed for forest departure
4. *rāja bhoga*:	midday meal
5. *utthāpana*:	wakes from midday rest
6. *sandhyā*:	twilight activities
7. *aulai darśana*:	evening meal at home
8. *śayana*:	retires for night

FIGURE 4. The Eight Periods of the Vraja-līlā

The eight periods of the meditative cycle and their respective eight *līlā* activities are diagrammed in Figure 4. The daily cycle commences with the beginning of night's end, a point which coincides with a moment called *brahma-muhūrta*. *Brahma-muhūrta* occurs three *muhūrta*s before sunrise (a *muhūrta* is a period of forty-eight minutes). It is held to be the most auspicious time of the day and is thus the time most serious Viaṣṇava practitioners rise from bed to begin their meditations. The meditative cycle follows each of the eight periods in order, until the cycle is completed and the day begins again. Each period has a particular *līlā*-event associated with it, and the practitioner is to visualize the appropriate event in the proper period. This is held to be an effective way of continually fixing the mind on and harmonizing it with the Ultimate Reality of Kṛṣṇa's play; in the spirit of the *Bhagavad-gītā,* the practitioner "goes to the state that he remembers." The aim is to perform this remembrance without interruption. Practical demands of the day do, of course, interrupt the meditations of the practitioner, at least until a very advanced stage. One *bābā* of Rādhākuṇḍa involved in this practice informed me that any activity which necessitates a break in the remembrance, such as cooking or sleeping, was either to be preceded or followed by a remembrance of that portion of the *līlā* missed while the mundane task was being executed.

Sources of the *Līlā:* Meditative Poetry

The tenth canto of the *Bhāgavata Purāṇa,* the Gauḍīya Vaiṣṇava scripture par excellence, describes the Vraja-līlā in relative detail. Over time a great amount of literature was produced which elaborated on this *līlā,* though of course the subsequent literature had to be in agreement with the *Bhāgavata Purāṇa.* Much of the subsequent literature was produced both from and for the *līlā-smaraṇa* meditation. The first description of the eightfold *līlā* was a short Sanskrit poem written by Rūpa Gosvāmin entitled the *Aṣṭa-kālīya-līlā-smaraṇa-maṅgala-stotraṃ* ("Auspicious Praise of the Remembrance of the Līlā Divided into Eight Time Periods") or simply the *Smaraṇa-maṅgala-stotram.*[33] Scholars in Vṛndāvana are of the opinion that this poem was largely inspired by the descriptions of the eternal *līlā* that appears in the Paṭala Khaṇḍa of the *Padma Purāṇa.* However, this claim is debatable. Niradprasād Nāth thinks that this section of the *Padma Purāṇa* must have been added after the time of the Vṛndāvana Gosvāmins, since it agrees so thoroughly with their theories yet is not quoted by them.[34] I am inclined to agree with Nāth. Regardless, the *Aṣṭa-kālīya-līlā-smaraṇa-maṅgala-stotraṃ* was the first text explicitly to divide the *līlā* into eight units. It provided a skeletal outline of the activities of each period which became the framework for later endless expansion. The poem consists of eleven verses. The first two praise Kṛṣṇa and his activities in Vraja. The meditation proper begins with the third verse, and the next seven verses go on to describe seven successive *līlā*-periods; the eighth time period, night's end, is the subject of the two closing verses. The description of the *līlā* of each time period is bracketed with the phrase "I remember . . ." (*smarāmi*). The cycle (diagrammed in Figure 4) is as follows.

At the beginning of night's end, Kṛṣṇa and his favorite lover Rādhā are awakened by birds sent by the forest goddess Vṛndā, and though still filled with passion, return to their separate homes fearful of the coming light. They are both bathed in the morning, then Rādhā goes to Kṛṣṇa's house at his father's village, Nandagrāma, and there feeds him. Kṛṣṇa then milks the cows. During forenoon, Kṛṣṇa goes forth from his village under the guise of tending the cows. Rādhā leaves her own village under the pretense of performing the sun worship, and the couple meet at the pond of Rādhākuṇḍa for hours of midday love-play. In the afternoon Rādhā and Kṛṣṇa return to their own villages, bathe, and prepare for the evening. Kṛṣṇa milks the cows again after sunset. In the late evening, Rādhā dresses for the night and sends a messenger to inform Kṛṣṇa of their meeting place in the forest of Vṛndāvana. Kṛṣṇa is entertained by the cowherds and then put to bed by his mother. He next sneaks out to the forest of Vṛndāvana to meet with Rādhā, and there they spend the night making love until sleep overcomes them. They are awakened at night's end and the daily cycle begins again.

The *Aṣṭa–kālīya-līlā-smaraṇa-maṅgala-stotraṃ* is visual poetry at its best; as mentioned before, it is important to understand that the poem was produced both *from līlā-smaraṇa* meditation and *for līlā-smaraṇa* meditation. In fact, if we place this poem in its particular context in the history of Gauḍīya Vaiṣṇava poetry in Vraja, we observe a chain of experience underlying all Gauḍīya Vaiṣṇava meditative poetry.

The Gauḍīya Vaiṣṇavas insist that the poet must be accomplished in *līlā-smaraṇa* meditation; all poetic embellishments are secondary to the direct meditative experience of Kṛṣṇa's *līlā*s. Since the author of the poem suddenly becomes very significant, a major feature of Gauḍīya Vaiṣṇava poetry is the signature line (*bhaṇita*) which validates the experience that the poem evokes.[35] The master meditator, Rūpa Gosvāmin, certainly fills the traditional requirement of meditative experience. From his meditations on Kṛṣṇa's *līlā*s, expressed in such scripture as the *Bhāgavata Purāṇa,* Rūpa experienced a vision. Interestingly, the *Bhāgavata Purāṇa* itself is considered to be the expression of a meditative vision. Having composed the *Brahma-sūtra* and brought the different Epics and Purāṇas into existence, the divine sage Vyāsa was still not satisfied; he therefore pursued a meditative course in which he obtained a direct vision of Ultimate Reality, which he then expressed as the *Bhāgavata Purāṇa*. As a result of his meditations on Kṛṣṇa's *līlā*s, as expressed in such Vaiṣṇava Purāṇas, Rūpa Gosvāmin acquired the vision he subsequently expressed in his poem, the *Smaraṇa-maṅgala-stotraṃ*. This poem, in turn, became the inspirational foundation of another, much longer Sanskrit poem (2488 verses) of great importance for the *līlā-smaraṇa* meditation: the *Govinda-līlāmṛta* of the influential Kṛṣṇadāsa Kavirāja.[36] As a result of his meditations on Rūpa's poem, Kṛṣṇadāsa was granted a vision of the *līlā,* which he then expressed in the poetic form of the *Govinda–līlāmṛta.* The verses of Rūpa's *smaraṇa-maṅgala-stotraṃ* are embedded within the *Govinda–līlāmṛta*'s descriptions of the activities of the eight periods. This poem gives a much more detailed account of the eightfold *līlā,* and itself became the inspirational foundation for yet later meditative experiences. The resultant visions and experiences yielded additional poetic expressions—the numerous Sanskrit *paddhati*s as well as the increasingly important Bengali *guṭikā*s.[37] The *Govinda–līlāmṛta,* or at least some rendition of this poem, is the text most frequently memorized and used as a basis of the *līlā-smaraṇa* visualization in Vraja today. Initiates must memorize all or at least significant portions of it; and although there are other important sources of the *līlā*s, these texts function significantly as the foundation for all later meditative experience.

A chain of poetic expression and meditative experience thus becomes observable. Poetry is used both to express the meditative experience and evoke the meditative experience. The process continues, generating more and more interdependent meditative experience and expression, expression and experience. Thus, a primordial revelation of the past becomes available to the practitioner of the present through meditative poetry and contemplative activity.

The interdependence of the meditative experience and poetry seriously challenges the view that religious experience, as Williams James so distinctly put it, "comes naked into the world."[38] Far from it. In this particular case, we can clearly observe how the meditative experience is the product of concentrated reflection on a religious text. The meditator participates in a world expressed within a text, or more precisely, within a poem.

Additional sources for knowledge of Kṛṣṇa's *līlā* include the stories of accomplished masters (*siddha*s) and finally the meditator's own direct vision (*sākṣāt-darśana*). The *siddha*s who have perfected their access to the world of the *līlā* become teachers for the less experienced practitioners. The resourceful *siddha* is often the practitioner's own *guru.* Much of the training in knowledge of the *līlā* comes from personal interaction with the *dīkṣā* and *śikṣā-gurus*, for an essential ingredient of the practitioner's initiation is the gradual revelation of increased esoteric knowledge of the *līlā*. A *siddha* is a treasure-house of *līlā* events, and stories told about the *līlā* adventures of famous *siddhas* become part of the sacred lore handed down over the years. Most of the stories remain oral, but occasionally a close disciple writes them down; these become central features of a *siddha*'s biography.

The practitioners must rely on the accounts of the *līlā* experiences of others for the basis of their meditations until they have perfected their own technique to the point where they can enter *samādhi* and see the *līlā* directly (*sākṣāt-darśana*). At this stage, one's own direct vision becomes the major source of knowledge of the *līlā*. The practitioner has now traveled far beyond the limited pointers of scripture and poetry, and has entered the land of direct experience where the *līlā* becomes infinite. Upon questioning a Vraja practitioner of *līlā-smaraṇa* meditation about the source of knowledge of the *līlā*, I received the following reply:

> Listen. The *līlā* is infinite. Texts like *Śrīmad Bhāgavatam* and *Govinda-līlāmṛta* are only examples of the *līlā*. In the beginning one needs the support of such texts as these, but when the *sādhana* becomes perfected one sees the *līlā* directly.

There is a point where one moves beyond paradigmatic action—where one must let go of all structure and plunge into the uncertainty of Kṛṣṇa's play. The practitioner has to learn some initial rules to get started, but finally this is a game without any rules. At the highest level of *līlā-smaraṇa* meditation, the mythic script leaps off the printed page and leads its actor into unpredictable eternal play.

Entering the Visualized *Līlā*

The practitioner of the Rāgānugā Bhakti Sādhana is not meant to stop at merely visualizing the *līlā;* he is to enter it as an active participant. The practitioner moves from a passive observer of the Vraja-līlā to an active participant by taking a part in that drama defined by the *siddha-rūpa*. Enacting this part implicates the practitioner directly in the world of the Vraja-līlā.

Meditative poetry, used to aid the process of visualizing the Vraja-līlā, also functions to pull the reader into that drama. Many Gauḍīya Vaiṣṇava poets produced poems that shift the reader's point of reference from that of observer to that of a participant in the world described. A poem by Narottama Dāsa serves as a good example.

> O Hari! Hari! When shall I attain that state
> in which I am born in Vraja as the daughter
> of a cowherd in Vṛṣabhānupūra?
>
> When shall I celebrate my marriage at Jāvaṭa
> and establish my residence there?
> When shall I serve the beloved of the dearest *sakhī*s? . . .
>
> I Narottama Dāsa reflect, when shall I become
> a female servant situated among
> the dearest *sakhī*s?[39]

This poem is clearly more than descriptive. These lines draw the reader into the world of the poem, causing the reader to assume the role of a *sakhī* or *mañjarī*. Thus the poems are vehicles of dramatic participation for the poet himself, as demonstrated by the signature line, and for the reader who experiences the implied role.

The stage onto which the practitioner of Mañjarī Sādhana initially moves is the *yoga-pīṭha,* graphically conceptualized as the previously described *maṇḍala*-like lotus. The *yoga-pīṭha* is a "place of union." It is a proper stage for Mañjarī Sādhana, since it is the place where Rādhā and Kṛṣṇa come together, and is also the occasion for the union of the *bhakta* and the divine couple. As one informant living in Vṛndāvana defined it: "The *yoga-pīṭha* is the place of union between Bhagavān and the *bhakta,* particularly the union of Bhagavān and Rādhā, the highest representative of the *bhakta*s." Since a *yoga-pīṭha* may occur anywhere in Vraja, potential *yoga-pītha*s are endless. However, there are three famous *yoga-pītha*s in Vraja, which every practitioner must know because of their central importance in the daily *līlā* cycle. The first is Kṛṣṇa's house in Nandagrāma, where Rādhā comes in the early morning to serve food to her beloved. The second is Rādhākuṇḍa, the pond where the divine couple meet at noontime for romance and sport in the water. And the third site is the forest of Vṛndāvana, where Rādhā and Kṛṣṇa rendezvous at night to embrace in a loving union. All of these meetings require the assistance of Rādhā's girlfriends, the *sakhī*s and *mañjarī*s with whom the practitioners identify.

The Rāgānugā practitioner must study the major *yoga-pīṭha*s and become familiar with their events, characters, environment, and other details. Initially, the practitioners use the meditative aid of the *maṇḍala*-like *yoga-pīṭhāmbuja* to visualize that portion of the Vraja-līlā which occurs at a particular site. The more advanced meditators no longer need the support of a diagrammed *yoga-pīṭhāmbuja.* Some practitioners even choose to realize a particular *līlā* event while sitting on the exact physical site where that event is believed to have eventually occurred. After the practitioner has visualized the *līlā* of a particular *yoga-pīṭha,* he enters it by taking the specific part defined by his own *siddha-rūpa.*[40] Every part involves a distinctive set of performative acts of loving devotion (*sevā*). For the *mañjarī,* this performance typically involves such acts of service for the couple as arranging trysts, providing entertainment, serving betel nut, fetching water, and so forth. At the initial stage the practitioner's performance is strictly defined and limited to a certain set of acts. The *guru* trains the practitioner in a few mechanical acts of service and carefully directs the performance in the cosmic drama. But as the *sādhana* is perfected and the practitioner enters the infinite dimensions of the *līlā,* his/her adventures become spontaneous, unpredictable, unlimited.

Though this type of activity is to take place on a meditative stage in the mind, all Gauḍīya Vaiṣṇava practitioners agree that some kind of performance with the physical body, the *sādhaka-rūpa,* is a necessary component of the *sādhana.* Action in one body influences developments in the other. Thus we now turn to the performance of the *sādhaka-rūpa.*

Performance with the *Sādhaka-rūpa*

There are really two interpretations of what form the performance with the practitioner's physical body should take, which correspond to the two conflicting interpretations discussed in Chapter 6. The first, following the interpretation of Viśvanātha Cakravartin and Kṛṣṇadāsa Kavirāja, asserts that the external performance consists primarily of commonly accepted devotional acts. The second insists that literal imitation of the *gopī*s is more appropriate.

The Standard Acts of *Bhakti*

In the *Caitanya–caritāmṛta,* Kṛṣṇadāsa Kavirāja defined the external form of the Rāgānugā Bhakti Sādhana with the *sādhaka-rūpa* as "*śravaṇa kīrtana,* etc."[41] These words bring to the mind of a Vaiṣṇava a list of nine types of *bhakti-sādhana* mentioned in the *Bhāgavata Purāṇa* (7.5.23). When asked to recite something he had learned from his teachers, Prahlāda replied: (1) *śravaṇa* ("listening" to the scriptural stories of

Kṛṣṇa and his companions), (2) *kīrtana* ("praising," usually refers to ecstatic group singing), (3) *smaraṇa* ("remembering" or fixing the mind on Viṣṇu), (4) *pāda-sevana* (rendering service), (5) *arcana* (worshipping an image), (6) *vandana* (paying homage), (7) *dāsya* (servitude), (8) *sākhya* (friendship), and (9) *ātma-nivedana* (complete surrender of the self). These words, placed in the mouth of the great Vaiṣṇava *bhakta* Prahlāda, come to be considered the very foundation of Vaiṣṇava *bhakti*. These acts of *bhakti* are incorporated as an integral part of the framework of Vaidhī Bhakti by Rūpa Gosvāmin, who states they also have a place in Rāgānugā (BRS 1.2.296). Many Gauḍīya Vaiṣṇavas will make the claim that Rāgānugā with the external *sādhaka-rūpa* consists solely of devotional acts such as those listed among the ninefold practice. Viśvanātha Cakravartin was of the opinion that Rāgānugā and Vaidhī Bhakti with the physical body differ not in form, but only in motive.

Moreover, as was shown, Viśvanātha maintained that Rāgānugā with the *sādhaka-rūpa* imitates Rūpa and Sanātana Gosvāmin, and the tradition certainly pictures the Vṛndāvana Gosvāmins as being deeply involved in such practices as *śravaṇa, kīrtana,* and the rest. More important, even some of those who opt to follow the exemplars of scripture maintain that these acts of *bhakti* were displayed by the paradigmatic individuals of the *Bhāgavata Purāṇa*. The ninefold practice was explicitly associated with the paradigmatic Prahlāda. But more relevant, the Gauḍīya Vaiṣṇavas see the roots of these acts of *bhakti* in the behavior of the *gopī*s themselves. When separated from Kṛṣṇa, the *gopī*s narrated and listened to the stories of Kṛṣṇa (*śravaṇa*), praised his glorious deeds (*kīrtana*), and performed other acts as a means of keeping him close to mind. Thus, whether one sees *śravaṇa, kīrtana,* and so forth, as exemplary acts of Rūpa and Sanātana or as exemplary acts of the *gopī*s, they are nevertheless considered inherent in the behavior of the paradigmatic Vrajaloka, and for many practitioners imitation of them constitutes Rāgānugā with the *sādhaka-rūpa*.

Kīrtana holds a prominent position in Gauḍīya Vaiṣṇava *sādhana* and deserves special attention for its dramatic possibilities. *Kīrtana* takes many forms, but typically it involves a communal gathering around a group of musicians. The musicians, accompanied by drums, cymbals, and flutes, are well trained to induce ecstatic emotional states in the audience. They sing poetic songs designed to draw the listener into the world they depict. I have attended Bengali *kīrtana* performances in Vṛndāvana in which many of the musicians and listeners were visibly moved to intense degrees. A favorite theme of the *kīrtana* sessions is the separation of the *gopīs* from Kṛṣṇa. At these sessions I observed several members of the audience exhibiting such actions as rolling on the ground, dancing wildly, and weeping uncontrollably—all classical *anubhāvas* displayed by the *gopīs* in the anguish of separation. The *kīrtana* performance, then, provides a means for the predisposed *bhakta* to participate in the emotional world of the *līlā* with the physical body.

Performance in the *Ṭhākura Ghara*

Another important type of practice, which also appears among the list of traditional Vaiṣṇava practices, is the worship of images (*arcana*). The maintenance and worship of images (*arcas* or *mūrtis*) is a common feature of Hinduism. But the Gauḍīya Vaiṣṇava relationship with and treatment of the images of Kṛṣṇa show a marked influence of the dynamics of the Rāgānugā Bhakti Sādhana. The care of an image among the Gauḍīya Vaiṣṇavas, viewed from the vantage point of our study, appears as another fascinating form of the performance with the *sādhaka-rūpa,* which results in the realization of a particular identity or relationship with Kṛṣṇa in the eternal play of Vraja.

Besides the many public temples of Vraja, there exists in the hut or home of many practicing *bhaktas* a space, frequently an entire room, where the individual deities or images (*mūrtis*) of the residents are kept. This room is called the "god-room," or *ṭhākura ghara.* For the Hindu, the image is a concrete form for concentration and the focus of many daily rituals. Worship and meditation require such a focus. The *Viṣṇudharmottara* establishes this point:

> The Supreme Self has two forms: *prakṛti* and *vikṛti. Prakṛti* is his invisible form. *Vikṛti* is the visible form in which he pervades the entire world. Worship and meditation can only be performed to the visible form (3.46.2–3).

Most Vaiṣṇavas agree that the most accessible and visible form in which the Supreme manifests itself is the *mūrti* or *arca.*[42]

However, the image is much more than a focus for worship and meditation for the Gauḍīya Vaiṣṇava; Kṛṣṇa actually embodies the image. Perhaps the word "image" is a poor translation of the Sanskrit term *mūrti;* "body" may be more appropriate. Regardless of which word we choose to translate *mūrti,* it is important to keep in mind that according to most Hindus, the formless takes form for the purpose of worship. God agrees to inhabit the *mūrti* out of love for his *bhaktas*. Once the image has been properly installed (*pratiṣṭhā*) with the appropriate rituals,[43] it is viewed as a body of God; this explains the extreme care and attention given to the images throughout India.

What makes the care of the image distinct among Gauḍīya Vaiṣṇavas in Vraja is the close personal relationship that is developed with the image, and the concomitant identity this nurtures in the worshipper. In Vraja, the image as Kṛṣṇa is held to be a master, child, friend, or lover of the worshipper. Images of Kṛṣṇa in Vraja are of two varieties: some are made of a special blend of metals and are fashioned in the form of the child or the adolescent; others are a stone taken from the Govardhana mountain.[44] The worship of the image roughly follows the ideal schedule of the temple, as indicated in the timetable of Figure 4.

Kṛṣṇa is awakened early in the morning, often from a miniature bed, bathed and fed, and then dressed for his outing in the forest. Throughout the day the image of Kṛṣṇa is served and prepared for the appropriate activity determined by the daily ritual cycle, until it is time to put him to bed once again. Women make clothes for the image. One German woman who had settled in Vṛndāvana and received initiation from a *guru* was instructed by her *guru* to tend an image of Kṛṣṇa in the form of the child. During the winter months she knitted little sweaters and ski caps for her image. Her *guru* was pleased, for this demonstrated a serious involvement with the image. The image is always treated as an honored guest. I have seen this in homes too poor to purchase more than one fan or heater: on a hot, sweltering summer day, the one fan was in the *ṭhākura ghara* cooling the images, while in the winter the family's single electric heater was placed in the *ṭhākura ghara* to warm the images. Much of the day is organized around care of the images. I observed a woman in Vṛndāvana bathe, feed, and clothe her baby Kṛṣṇa, then put it in a cradle and tend it for hours. The images should never be left unattended. A family with whom I frequently attended festivals would always leave one member behind to care for the needs of the images. On Kṛṣṇa's birthday, toys are placed before the image and are set in motion for his entertainment. An outsider witnessing these activities is reminded of doll-playing. I do not think the Gauḍīya Vaiṣṇavas would greatly object to this oversimplified comparison, for children often experience real feelings for and attachment to their dolls. "Doll-playing" with the installed images, however, is a much more serious affair; its goal is to generate a loving relationship with and attachment to Kṛṣṇa, as he is present in the image.

What is perhaps most interesting about the activities in the *ṭhākura ghara* from our perspective is their transformative effect. Rūpa mentions an old man who achieved the state of perfection (*siddha,* understood to be the state after permanent transformation of identity has been achieved) by worshipping an image of Kṛṣṇa as his son (BRS 1.2.307). Anticipating a relationship with Kṛṣṇa necessarily causes one to assume the identity of a concomitant role. Even in secular doll-playing, the child frequently assumes an identity other than the ordinary one—e.g., the doll's mother. So, too, the "doll-playing" of the *ṭhākura ghara* places the practitioner in a very particular role vis-à-vis Kṛṣṇa. A shift in the identity of the worshipper occurs as he or she enters the *ṭhākura ghara.* Several informants told me that they never enter the *ṭhākura ghura* to serve Kṛṣṇa as the person I perceived. Rather, they enter the *ṭhākura ghara* in the identity of their *siddha-rūpa.* (Though of course the activities therein are carried out with the physical body.) One informant residing in Vṛndāvana made the following remarks:

> I am another person in that other world. I never enter the *thākura ghara* and worship the deities as you see me, I always enter as a *mañjarī,* for only in this form can one enter the eternal Vṛndāvana. In the morning, I bathe Kṛṣṇa in my mind with this form as I bathe my Govardhana stone with this physical body.

Interaction with the image is a way of pursuing the identity revealed by the *guru,* a means of enacting the given role. The *ṭhākura ghara,* then, constitutes a ritual stage on which the practitioners can physically act out their roles in the *līlā* every day in their own homes. The dramatic process of Rāgānugā thus underlies and informs the daily image worship of many Gauḍīya Vaiṣṇavas, providing a means to participate in the world of the Vraja-līlā even for those who are unable to practice the more difficult meditative techniques.

Physical Role-Taking

Though the practice is condemned by most orthodox Gauḍīya Vaiṣṇavas, one still can find male *bābā*s in the vicinity of Vraja dressed as females and following female habits.[45] When asked why they do this they will typically respond that, though they appear to be male, they are really female and should therefore dress accordingly. One may also receive the reply that this is part of their *sādhana.* This practice, which may seem ludicrous to the outsider, is entirely consistent with the direction of the Rāgānugā Bhakti Sādhana and represents one interpretive strategy for imitating the Vrajaloka with the physical body. It is an extreme attempt to identify with the true inner and essential nature (*siddha-svarūpa*), which is usually conceived of as a female *gopī.* It is a physical effort on the part of some practitioners to transform the identity from its location in the ordinary body to the ultimately real body as revealed by the *guru,* and thereby inhabit the mythical world of Vraja.[46] As Stanislavski has taught us, outer physical acts lead to the inner world of a character.

This technique has been used as a transformative device by others in India who have been influenced by the Gauḍīya Vaiṣṇava theory of *sādhana.* The nineteenth-century Bengali saint Ramakrishna, for example, dressed as a woman for a period as part of his *sādhana.* Closer yet, in his early study *The Religious Sects of the Hindus,* H. H. Wilson reports of a group called the Sakhī Bhāvas, found throughout Vraja, Rajasthan, and Bengal.

> In order to convey the idea of being as it were her (Rādhā's) followers and friends, a character obviously incompatible with the difference of sex, they assume the female garb, and adopt not only the dress and ornaments, but the manners and occupations of women.[47]

David Kinsley reports of a similar Gauḍīya Vaiṣṇava movement in the Calcutta area and again associates it with the Sakhī Bhāvas.

> Another group, most likely direct descendants of the orthodox Bengali Vaiṣṇavas, was becoming popular in Calcutta. The members of this group believed themselves to be *gopī*s (a common-enough belief among orthodox Vaiṣṇavas) and dressed the part. Unlike the orthodox Bengali Vaiṣṇavas, who condemn such practices, this group, called the Sakhībhāvaks, dressed as women in imitation of the *gopī*s as a regular part of their devotional *sādhana.* Some of the members, particularly those who went to live at Vṛndāvana, wore feminine dress throughout their lives.[48]

These practices have aroused the scorn of modern Hindu reformers, who have little knowledge of or sympathy for the theory which underlies the practice. R. G. Bhandarkar, for example, writes:

> The worship of Rādhā, more prominently even than that of Kṛṣṇa, has given rise to a sect, the members of which assume the garb of women with all their ordinary manners and affect to be subject even to their monthly sickness. Their appearance and acts are so disgusting that they do not show themselves very much in public, and their number is small. Their goal is the realization of the position of female companions and attendants of Rādhā; and hence they probably assume the name Sakhībhāvas (literally, the condition of companions). They deserve notice here only to show that, when the female element is idolized and made the object of special worship, such disgusting corruptions must ensue.[49]

Because of such strong criticism, *bābās* who interpret Rūpa's instructions to mean that one is even physically to imitate the *gopīs* tend to maintain a very low profile. They keep to the more remote areas of Vraja, appearing only at such important festivals as the celebration of Rādhā's birthday. Nevertheless, practitioners of this type do exist in Vraja today and are usually associated with either the Gauḍīya Vaiṣṇavas or the Rādhā Vallabhīs, a sect closely related to the Gauḍīyas. At the turn of the century there was a famous and well-accepted Gauḍīya Vaiṣṇava *bābā* who went by the name of Lalitā Sakhī and passed his/her life dressed as a female.[50] As the name suggests, this *bābā* strove to realize the role of this important *sakhī* of the Vraja-līlā by physically dressing and acting the part. As long as these practitioners remain few and mostly hidden from the public eye, they are generally tolerated and even occasionally hailed as rare saints. The orthodox are much more concerned with denying the validity of males imitating Kṛṣṇa and acting out his love affairs with female practitioners.[51]

Though the imitation of a *gopī* in any explicit or extreme form is generally avoided in Gauḍīya Vaiṣṇava *sādhana,* symbolic acts of such role-taking are quite acceptable. One frequently sees a *kīrtana* leader decorated with an ornament or a piece of clothing of a woman, such as a shawl, as he takes the part of a *gopī* in a particular song. Some of the *bābās* will paint the bottoms of their feet red (a mark of a *gopī*), which is another way of symbolically indicating that they are really women. Others may tell you that the piece of cloth they wear is a feminine garment. It is not uncommon to hear a male Hindi-speaking (Bengali does not indicate gender) *bābā* using feminine grammatical endings when referring to himself, an additional means of stating womanhood. A parallel example is observable in the temple worship of Vallabhācārya Sampradāya, a sect which commonly follows the emotional role of the mother of Kṛṣṇa. The Gosvāmins of this sect always wear a bangle on their wrist when performing the public worship of the temple image, a symbolic connection with their exemplary role. These acts are a symbolic assent to the intent of the more explicit practice of imitating the *gopīs* in all ways, including dress and manners. Though they are of subtler form, they too reveal dramatic behavior designed to realize a role whose identity is enmeshed in the world of the Vraja-līlā.

Pilgrimage in Vraja

One of Rūpa's most important instructions for the Rāgānugā practitioner is to live continually in Vraja in a state of attachment to the various stories of Kṛṣṇa (BRS 1.2.294). All commentaries on this verse stress that this should be done physically if at all possible. Several Gauḍīya Vaiṣṇavas living in Vṛndāvana told me that the most important *sādhana* they were performing was simply living in Vraja. Some of the older *bābās* take a vow never to leave the region of Vraja for the rest of their lives. Thus, simply being physically present in the land of Vraja has religious significance for the Gauḍīya Vaiṣṇavas.[52] Why? In his *Upadeśāmṛta,* Rūpa writes:

> Madhupurī [Mathurā] is superior to Vaikuṇṭha because of [Kṛṣṇa's] birth there. Vṛndāraṇya [Vṛndāvana] is even better because of the celebration of the *rāsa-līlā* there. Govardhana is better yet since there [Kṛṣṇa] delighted all by raising it with his hand. And Rādhākuṇḍa is best of all, because of the sprinkling here of the nectar of love of the Lord of Gokula.[53]

We see from these lines that the sites within Vraja are important because they are associated with some activity of Kṛṣṇa. The natural phenomena of Vraja—its rivers, rocks, ponds, hills, woods, and even its soil—are objects of religious attention because of their contact with Kṛṣṇa at the time of the manifest *līlā*. Vraja is the stage on which Kṛṣṇa and his companions once performed the original drama long ago. (It still continues there in an unmanifest form [*aprakaṭa*] for those whose eyes have been opened by meditation.) Therefore, Vraja is the stage on which the Gauḍīya Vaiṣṇava practitioners can most perfectly follow in the footsteps of the Vrajaloka and act out their salvific roles. Rūpa quotes a verse from the *Padma Purāṇa* which states that if one resides in the Mathurā-maṇḍala[54] for just one day, *bhakti* for Lord Hari is produced (BRS 1.2.237). Such statements are taken seriously by most Gauḍīya Vaiṣṇavas today. All Gauḍīya Vaiṣṇavas try to visit Vraja at least once in their lifetime. Many choose to retire there for more permanent religious practice. This concentrated interest in Vraja has caused it to be one of the most important pilgrimage centers in northern India,[55] and the Gauḍīya Vaiṣṇavas have been extremely instrumental in the cultural development of the area.

The requirement that one should dwell in Vraja—at least for a short time if not permanently—and that one should remain intent upon the stories of Kṛṣṇa, has produced a distinctive style of pilgrimage in Vraja. Toward the end of the rainy season, thousands of pilgrims flock to Vraja to participate in a pilgrimage known as the Vraja-*yātrā* or *vana-yātrā* (modern Bengali *bon-jātrā;* literally, "procession through the forests").[56] Charlotte Vaudeville makes the claim that "the great authority for the Braj-parikrama or Ban-yatra is considered to be the *Mathurā-māhātmyam,* a religious chronicle of Mathurā found in the *Vārāha Purāṇa.*"[57] Some local scholars of Vraja see the prototype of the *vana-yātrā* in the activities of Uddhava when he visited Vraja to console the love-torn *gopīs.*[58] Most scholars, however, trace the development of the *vana-yātrā* to the career of Nārāyana Bhaṭṭa.[59]

Nārāyana Bhaṭṭa was a contemporary of the Vṛndāvana Gosvāmins. Local legend has it that he was an incarnation of Nārada.[60] One wonders why he did not become known as one of the Vṛndāvana Gosvāmins; his writings certainly merit attention.[61] A Sanskrit biography, written toward the end of the seventeenth century by one Jānkīprasāda,[62] informs us that Nārāyana Bhaṭṭa was born in Madurai in 1531. He had an early fascination for Vraja and soon took up residence at Rādhākuṇḍa. There, after receiving initiation into the Gauḍīya Vaiṣṇava Sampradāya from a *guru* named Kṛṣṇadāsa Brahmacārin, a disciple of Gadādhara, he wrote seven books which reveal his great interest in Vraja: *The Manifestation of Bhakti in Vraja (Vraja-bhakti-vilāsa), The Light of Vraja (Vraja-pradīpikā), Moonlight on the Festivals of Vraja (Vrajotsava-candrikā), The Great Ocean of Vraja (Vraja-mahodadhi), The Joy of the Celebrations in Vraja (Vrajotsavāhlādinī), The Great Celebration of the Qualities of Vraja* (*Bṛhat-Vraja-guṇotsava*), and *The Illumination of Vraja* (*Vraja-prakāśa*). After his residence at Rādhākuṇḍa, Nārāyana Bahṭṭa established a temple at Barsana, Rādhā's village, and lived nearby writing further books on *bhakti.* Perhaps no one did more than Nārāyana Bhaṭṭa to work out the correlation between the mythological stories of Kṛṣṇa and his companions and the physical sites of Vraja. He was very active in establishing many of the chief *līlā* sites that became the focus of the *vana-yātrā.* Moreover, it is most probable that Nārāyana Bhaṭṭa was the first to develop the distinctive dramatic pilgrimage of Vraja, the *vana-yātrā.*

The *vana-yātrā* pilgrimage lasts an average of thirty days and consists of visiting the sites of the various *līlā*s of Kṛṣṇa and his companions scattered throughout Vraja, and witnessing there the performance of a type of drama known as the *rāsa-līlā.*[63] Occasionally the pilgrims even take the part of the Vrajaloka themselves in the dramatic performance acted out on the original "stage." The *rāsa-līlā* is a religious drama that could be seen as an example of the somewhat common genre identified by Northrop Frye as the "scriptural play" or "myth play," since the scripture or myth defining the ideal world of the tradition forms the script of the play.[64] As such, it functions as both a model of and model for the tradition's religious reality.[65] The script of the *rāsa-līlā* is paradigmatically rooted in the *Bhāgavata Purāṇa* and other authoritative Vaiṣṇava scriptures and, as such, the *rāsa-līlā*s make vividly present the ideal religious world of Vraja as well as the paradigmatic roles that enable one to achieve that world. The *rāsa-līlā* hence serves a function similar to that of the *līlā-smaraṇa* meditation; it provides visual access to the Vraja-līlā. All this takes place on the very "stage" where the original event is believed to have occurred. Regarding the ordinary staged *rāsa-līlā,* Norvin Hein writes: "Through the play the Hindu . . . re-enters the world of traditional religious truth. There he dwells in the loving emotion which establishes and confirms for the Vaiṣṇava, his positive orientation toward the universe."[66] How much more, then, is this true for the *rāsa-līlā*s of the *vana-yātrā,* which take place not on any ordinary stage but in the groves of Vraja. Hein has well described this unique convergence of drama and pilgrimage.

> The most extensive program of *rāslīlā* known in Braja is a series which is staged as part of observances of the *banjātrā,* which is a many-day tour through the countryside on a meandering trail about 120 miles in length. It includes visits to almost every wood and grove, every pond and cave that legend associates with the combats, antics, and loves of the boy Krishna. . . . Each evening while on the trail, the multitude of pilgrims gathers to see repeated, by svarūps who are for the time being the deities themselves, the very deeds which were done on the soil on which they sit.[67]

The *rāsa-līlā*s of the *vana-yātrā* are frequently performed on a small circular stage called a *rāsa-maṇḍala.* The *rāsa-maṇḍala* is theoretically located on the exact spot on which the *līlā* being celebrated was originally enacted. Frequently, it stands alone in the forest, wall-less and open to the sky, although sometimes connected to a temple. Regardless, these places form the main focus of the *vana-yātrā* and mark the stations of the pilgrim's trail that winds through the mythical land of Vraja. It might be important to add that these sites, thus marked, also provide a "stage" for some of the *bābā*s engaged in the *līlā-smaraṇa* meditation to perform their mental meditative dramas.

As the practitioners of the *līlā-smaraṇa* meditation eventually move from the position of a passive observer to an active participant in the Vraja-lila, so too does the *vana-yātrā* pilgrim. This pilgrim is not satisfied with watching a *rāsa-līlā* performance on an isolated stage in some distant land, but rather desires to travel to the original site to witness the enactment of the *līlā* and somehow participate in that dramatic world. In Vraja the pilgrims have increased opportunity to assume the roles of the

Vrajaloka portrayed within the *rāsa-līlā*s, and the *vana-yātrā* pilgrimage provides them with the environment to live out those paradigmatic roles. What better place could there be to imitate the original inhabitants of Vraja than Vraja itself? Like those original inhabitants, the pilgrims can experience directly the call of the peacock at sunset; bathe on the banks of the Yamunā; move around Govardhana mountain; and walk in the dust of Vṛndāvana, made holy by the contact with Kṛṣṇa and his intimates. The ideal pilgrims try to imitate the Vrajaloka throughout their travels. They travel about in wagons believed to be identical to the wagons of the original inhabitants of Vraja. They bathe in the pond in which Śiva bathed to attain the body of a woman. They tie bits of cloth to the very tree in which Kṛṣṇa hid the clothes of the *gopī*s. As mentioned earlier, they even occasionally assume the roles of the Vrajaloka in the *rāsa-līlā* performances. But most important, many pilgrims consciously assume the emotional role of a *gopī* in love with Kṛṣṇa and enter his play in a number of ways.

Observable in the playful activities of the *vana-yātrā* is a clear attempt to "live a myth" through physical action. The entire world of the mythical Vraja-līlā has been superimposed on the area now known as Vraja, identifying it as the very place where Kṛṣṇa and his companions originally enacted their *līlā* long ago and still continue to do so for those initiated into the ways of Rāgānugā. The pilgrim in Vraja is an actor on the ideal stage and through role-identification is in the very same position as the *gopī*s and other intimate companions of Kṛṣṇa, who first animated the land of Vraja. All things that were can now be, for the one who enters the mythical world of Vraja by assuming the part of one of its inhabitants.

These, then, are some of the performative acts of Gauḍīya Vaiṣṇava practitioners. What they all have in common is the underlying principle of identity transformation, pursued by assuming a particular role located within the world of the Vraja-līlā. Many householders do not have time to learn or practice the more rigorous *līlā-smaraṇa* meditation and perform only the outer forms of *sādhana.* But even the temporary ritual activities of the *vana-yātrā* pilgrim will have a lasting effect after the pilgrim leaves Vraja. As Geertz says: "Having ritually 'lept' into the framework of meaning which religious conceptions define, and the ritual ended, returned again to the common-sense world, a man is—unless as sometimes happens, the experience fails to register—changed."[68] This change is even greater for the more advanced practitioner who daily assumes the role of one of the special Vrajaloka—through worship in the *ṭhākura ghara,* through the more extreme path of physical role-taking, or especially through the meditative experiences in the *līlā-smaraṇa* meditation. Though the two forms of *sādhana* are usually practiced together, the inner meditative performance is held to be superior to the outer performance since it makes use of the *siddha-rūpa,* which has direct access to the *līlā,* whereas the outer utilizes the illusion-bound *sādhaka-rūpa.* Yet the outer physical practice supports and influences the inward development and is therefore judged to be highly beneficial, at least until the highest stage has been reached. At the highest stage, when the practitioner is continually absorbed in the *līlā,* several *bābā*s told me the outer forms of practice, such as service of an image, could be given up. This highest stage of perfection (*siddha*) occurs when total identification with the *siddha-rūpa* has been achieved. Such permanent absorption (*āveśa*) in the limitless Vraja-līlā, we have seen, is the goal of the *sādhana.*

The Rāgānugā Bhakti Sādhana begins as a mechanical imitation of the spontaneous external expressions, or *anubhāva*s, of a general role exemplified by one of the perfected Vrajaloka, with the belief that this procedure can reproduce the experience of that exemplary character. The performance is first based on a limited knowledge of the *līlā* as expressed in scripture and poetry. As the *sādhana* develops, the practitioner gradually refines the general role to a more particularized one, as the identity of the *siddha-rūpa* is consciously imposed. This process continues until the *sādhana* is perfected and the practitioner makes the shift to actually becoming the character defined as the *siddha-rūpa;* then the enactment of the role becomes spontaneous. At this point, direct access to the *līlā* has been achieved, and the practitioner's actions are no longer conscious and mechanical—*sādhana*—but are the free expressions of true inner experience—*anubhāva.* We witness here the return to the *anubhāva,* which is the proof for the tradition that a practitioner has reached the perfect state and participates fully in the Ultimate Reality of the Vraja-līlā.

8

On to Different Stages

Living a myth . . . implies a genuinely religious experience.
MIRCEA ELIADE, *Myth and Reality*

One final act remains before the curtain falls on this *līlā*. In this chapter I briefly summarize the central points of the thesis, and then coax its implications off the particular stage of Gauḍīya Vaiṣṇavism to suggest performances on the stages of different traditions, thereby demonstrating the comparative nature of the type of religious action we have observed in our study of the Rāgānugā Bhakti Sādhana. As this is not the occasion for a complete performance on any other stage, the application of the thesis to the context of different traditions must remain somewhat speculative, but I hope enough will be said to make the plausibility of the proposed comparative model evident.

Concluding Remarks

This study is intended to stand as a conscious contradiction and challenge to those who claim that means and methods have no place in Hindu *bhakti*. It should now be quite clear that *sādhana*, an intentional method or technique designed to realize the ultimate goal, does occupy a prominent position in the type of religion known as Hindu *bhakti*. Indeed, it must be admitted that certain schools of *bhakti* do place heavy, even exclusive, emphasis on unmerited grace.[1] It is important to note, however, that in these particular schools of *bhakti* the theologians usually prefer the term *prapatti* (total surrender) to the term *bhakti*, which they assume to involve intentional techniques, such as meditation, to define their distinctive religious position.[2] Perhaps it would be useful to differentiate between the Hindu *bhakta*, who favors *sādhana*, and the

Hindu *prapanna,* who favors unmerited grace. Regardless, *sādhana* does feature importantly in Hindu *bhakti.*

The recognition of this fact is a crucial one for this study, since I have used one important school of Hindu *bhakti,* Gauḍīya Vaiṣṇavism, to demonstrate a path of salvation that employs *methods* of dramatic imitation. I do not claim that this type of religious action covers all religious action; nevertheless, it is an important one, and one to be found in many religious traditions. It is generally utilized by those who value and believe in the need for human effort in the process of salvation, though it is rarely independent of notions of divine grace. This type of religious action is also frequently associated with followers of "monastic" forms of religions, since the intense pursuit of an alternative identity usually requires some kind of withdrawal from the ordinary business of society.

Although the idea of unmerited grace is certainly acknowledged as valid among early Gauḍīya Vaiṣṇavas and appears to gain ground over time, Rūpa Gosvāmin maintains that salvation by grace alone is extremely rare[3] and, therefore, that most do not achieve the ultimate goal without *sādhana.*[4] The *sādhana* he recommends is, of course, the Rāgānugā Bhakti Sādhana, which I have presented as a ritual process formulated with dramatic theory, designed to transform the practitioner's identity in order to lead him into the ultimate religious world of the Vraja-līlā.[5]

The chief concern of this study was to determine how individuals come to inhabit particular religious realities. Close attention was given to identity formation. Indeed, the entire process of entering the Ultimate Reality of Vraja is heavily dependent on the assumption of an identity defined by a paradigmatic figure who inhabits that mythological world. Religious experience is linked to the constitution of a specialized identity. An observable process of religious identity formation emerges from a study of the Rāgānugā Bhakti Sādhana. The process can briefly be summarized as follows.

The process begins with an exposure to a particular definition of reality or religious "script," which is made objectively available through various media. Scriptural readings, poetic recitals, storytelling, and dramatic presentations[6] are popular ways for the Vaiṣṇavas to make the world of Vraja, complete with its exemplary roles, objectively available. If the encounters with these media are effective as aesthetic experience, the audience identifies temporarily with the world of the depicted characters and the religious script opens up as possibility. Concomitant with this temporary identification is an implicit temporary "depersonalization"—

that is, as one embraces a new identity, even temporarily, one dissociates oneself from one's previous identity. Sometimes the experience ends here and remains purely "aesthetic."

If, however, the "script" becomes laden with an authority that is accepted to be more real, meaningful, and beneficial than one's present reality, there arises the desire to enter and participate permanently in that depicted reality, the locus of real meaningful existence. At this point, one has crossed the border of purely aesthetic experience and entered the dimension of religion. When a particular world is accepted as ultimately real from an objective standpoint, one is faced with the problem of defining "how to get there." This is where the process seems to begin for Rūpa. He spends no time arguing for the acceptance of the definition of reality presented in the *Bhāgavata Purāṇa*. Instead, he concerns himself with the ways and means with which to enter that world.

The objectified "script" contains ideal, exemplary roles or paradigmatic individuals which are the guides into the desired reality, for as the paradigmatic individuals are located firmly within that world, the identities they present become the very vehicles into that world. By following the procedures of Rūpa's Vaidhī Bhakti, the Gauḍīya Vaiṣṇava practitioner is exposed to and learns the exemplary roles. Next, the self-identity of the practitioner progressively and more permanently begins to shift to one of the depicted roles. In the Rāgānugā Bhakti Sādhana, the practitioner imitates the emotional roles (*bhāva*s) displayed by the original characters of Vraja, the Vrajaloka. A progressive and more permanent process of depersonalization occurs simultaneously with the assumption of the new identity. We have seen that the practitioners of the Rāgānugā Bhakti Sādhana gradually shift their identity from the one given in the ordinary process of socialization (*taṭastha-rūpa* or *sādhaka-rūpa abhimāna*) to an identity revealed by the *guru* which locates the practitioner in the world of Vraja (*siddha-rūpa abhimāna*). The degree to which the older established identity is left behind (depersonalization) is proportional to the degree the new identity is assumed, and vice versa.

The process of identity transformation presumably continues until total identification with the ideal role—our definition of "salvation"—has been achieved. Once this occurs, the world in which that role is located becomes subjectively real. For the Gauḍīya Vaiṣṇavas, this amounts to the realization of the emotional state of a *gopī* in the ultimate world of Vraja.

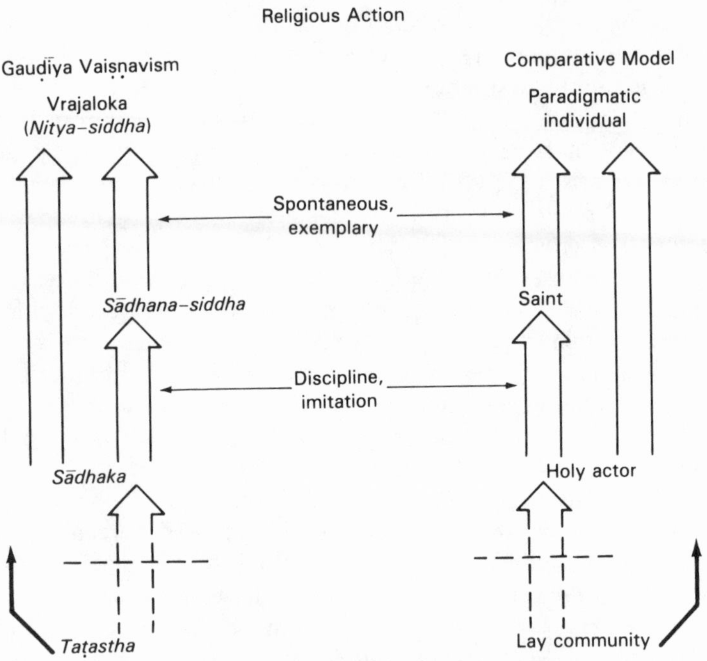

FIGURE 5. Comparative Model of Religious Action

A Comparative Model

Use of this imitative process, which aims at achieving the ultimate religious world of a paradigmatic individual, is not restricted to Gauḍīya Vaiṣṇavism. The Rāgānugā Bhakti Sādhana is an example of a structure of religious action common to a number of religious traditions. In conclusion, I propose to outline a cross-cultural model that can perhaps best be described as a path of religious action traveled by "holy actors."[7] To demonstrate the comparative nature of this general model I will make brief references to three other religious traditions: Cistercian Christianity, Theravāda Buddhism of Southeast Asia, and the religion of the Sioux visionary Black Elk.

While this comparative structure is being outlined, it will be useful to refer to the diagram of Figure 5. This diagram is based on insights that came to me while lying in the warmth of my courtyard in Vṛndāvana, looking up at the clear blue winter sky of northern India. High above me, I noticed a rather large bird floating effortlessly in the air. Suddenly two kites came into my vision. (Kite-flying is a favorite pastime of many inhabitants of Vraja.) The kites tried so hard to fly like the bird above; they strained and struggled to gain altitude. It dawned on me that this was an apt analogy for the three positions of Hindu aspiration: the *siddha,* the *sādhaka,* and the *taṭastha.* The bird flying effortlessly high above was the *siddha,* the perfected one, either an eternally perfected one (*nitya-siddha*) or one who had reached perfection by means of spiritual practices (*sādhana-siddha*). The bird, as *siddha,* presented a model of perfection for the kites to follow. The kites, then, were the *sādhaka*s, those aspiring to attain the world of the bird through imitative behavior. (The kites of Vraja even resemble birds.) I, the observer, occupied the last position, the *taṭastha,* as one carefully examining the activity from the outside, not yet giving effort to flight myself, but somehow benefiting from the performance which gave a sense of flight. This structure is illustrated in Figure 5.

At the top of the diagram is the "paradigmatic individual," a *superhuman* being whose exemplary life is revealed in the myth as a model of perfection. The paradigmatic individual has a double nature, divine and human. The divine nature of the paradigmatic individuals makes them worthy of being imitated, and the human nature makes imitation of the paradigmatic individuals by humans possible. We have seen that the Vrajaloka of the Gauḍīya Vaiṣṇavas are good examples of paradigmatic individuals. The Vaiṣṇava scriptures are full of their exemplary lives. They are held to be models of perfection because they have attained Kṛṣṇa (*Kṛṣṇa-prāpta*), and their divine nature is expressed by the fact that they are manifestations of the essential nature of Kṛṣṇa (*svarūpa-śakti*). Yet they are also viewed as having human form (*mānuṣa-rūpa*) and display salvific emotions available to humans. Personal models of perfection present the most accessible form of religious ideals.

The dual-natured paradigmatic individual, however, is also found in other traditions. Christ is such an exemplary figure for the Cistercians. Belief in the two natures of Christ is an important feature of Cistercian theology. Christ, as revealed in the Gospels, is the perfect Image and Likeness of God, to which humans are to conform in order to regain their lost likeness.[8] But he is also, the Cistercians insist, fully human, and as such is a suitable model for humans. The Buddha of the Pali Canon functions as a paradigmatic individual for the Theravāda monks of Southeast Asia. He was born in a pure and unusual way and, as the first to achieve *nirvāṇa,* was the Pathfinder. Buddha was also quite human, according to the Theravādins; what he attained we humans can also attain. Another interesting example of a paradigmatic individual is the sacred Red Man, revealed to Black Elk in his childhood vision.[9] The Red Man of Black Elk's vision is part bison and part man; it is he who revives the withering tree at the center of the Sioux nation and causes it to bloom with new vitality. Throughout his life, Black Elk strove to identify with this paradigmatic figure and embody his sacred power. The sacred Red Man, Buddha, Christ, and the Vrajaloka—all function similarly in their respective traditions. (The pluralistic nature of the Hindu tradition is highlighted by the fact that the Vrajaloka comprise a variety, whereas the paradigmatic individuals of many other traditions are singular.) These paradigmatic individuals represent perfection with their divine nature, and with their human nature demonstrate the human pathway to that perfection. The exemplary roles displayed by the paradigmatic individuals are the vehicles into the mythical worlds which they inhabit. The ultimate worlds indicated by the various paradigmatic individuals may be quite different, but the *means* of attaining those worlds is often the same: imitation.

Near the bottom of Figure 5 is the "holy actor." The holy actor is a religious practitioner who strives to enter the Ultimate Reality—or "live a myth"—by enacting a transcendent role defined by a paradigmatic individual. The holy actor is frequently a monk-like character, since serious pursuit of the ideal identity usually requires some type of renunciation of the ordinary concerns of society. All the religious action of the holy actor is informed by a discipline aimed at the realization of Ultimate Reality through imitation of the paradigmatic individual. The present study has been an attempt to demonstrate that the Rāgānugā Bhakti Sādhana is such an imitative process.

This imitative process is found elsewhere. The imitation of Christ informs the entire life of a Cistercian monk.[10] The Cistercian monk is a holy actor who strives continually to "conform" himself to the "form" of Christ by means of monastic discipline. The monastery is a controlled environment, a stage on which the monk can enact, under the directorship of an abbot, the role revealed by Christ. The Theravāda *bhikṣu* is a holy actor who removes himself from the ordinary concerns of society to follow in the footsteps of Buddha. Upon entering the monastery, the *bhikṣu* dramatically reenacts the "great retirement" of Buddha.[11] The *bhikṣu*'s subsequent life within the monastery is patterned after Buddha's *dharma* by means of discipline (*vinaya*).[12] The Sioux believed that the world revealed in a vision could be made present only if the vision was physically reenacted.[13] Therefore, Black Elk organized, with the help of a dream cult, a communal performance based on the script of his vision. During this communal performance Black Elk himself, using costume and paint, dramatically enacted the role of the sacred Red Man and ritually renewed the Tree of Life at the center of the Sioux nation.[14] Black Elk tried to "live a myth" literally and participate in its sacred reality; others have attempted to do so symbolically. But in each of these examples we observe a common structure: the religious participants strive to enter and participate in the ultimate world of meaning by imitating the paradigmatic individual and thereby internalizing a transcendent role.

The role of the paradigmatic individual is located in scripture for the first three traditions and in a vision for Black Elk. The act of transforming the role of the paradigmatic individual from a literary possibility to a physically embodied actuality requires much effort. This work of the holy actor is very much like that of an ordinary actor learning a part from a written script. Constantin Stanislavski aptly describes the process of preparing a role for the actor who is concerned with the "creative process of living and experiencing a role." He divides the preparatory work into three main periods: studying the role, establishing the life of the role, and putting the role into physical form.[15]

Under the Stanislavski Method, the studied reading of the script is a very important step in the creative process of preparing a role. This is where the actors first encounter the role they are to enact, which, if they are sensitive, immediately begins to influence them. The actors should read the script with an "open soul," with free feelings, as they temporarily identify with the role in their own minds and are "carried away by the reading." Moreover, Stanislavski says, "there are plays whose spiritual essence is so deeply embedded that it takes great effort to dig it out."[16] Thus the actor must meditate on the text and visualize the scenes with much time and care.

Similarly, meditative techniques of visualization based on scriptural scenes are a regular feature of the practices of the holy actor. We have seen how the meditative practice of *līlā–smaraṇa* involves visualizing the mythical world of Vraja, its various scenes, and its paradigmatic roles. Likewise, the Cistercian abbots stressed the importance of a practice called *lectio divina,* careful meditative reading and visualization of the monastic script, the Bible. This practice is important because, through it, the monk gains imaginary access to the ideal religious world and therein encounters the role he is to enact, namely Christ. The Theravāda *bhikṣu* is to reflect continually on the life and *dharma* of the Buddha, and visualization techniques of *smaraṇa* also have a place in the

Theravāda tradition.[17] Black Elk spent much of his time reflecting on his vision and trying to keep the key role of the sacred Red Man present in his mind.

From this early stage, Stanislavski encourages the actors to participate imaginatively in the script "as if" they were present there; that is, they are to enter the script by imaginatively identifying with a character or role located within the script. This is also an important feature of the practice of *smaraṇa* and *lectio divina,* in which the practitioner begins to experience aesthetically the world depicted in the scripture by imaginatively identifying with the paradigmatic individual.

But the Method actor seeks more than a private imaginative reading of a script, and the holy actor certainly seeks more than a temporary aesthetic reading of scripture. Both seek physically to embody the role they have encountered therein. Before this can be done, however, the actors need to analyze and understand thoroughly the character they are to enact. This involves an analysis of how the character thinks, feels, and acts in all situations. We saw that a careful analysis of the roles of the various types of Vrajaloka was a major concern of Rūpa Gosvāmin. Likewise, Christology and Buddhology are a necessary aspect of the Christian and Buddhist monk's knowledge.

The third step of the Method actor's preparation for a role is what Stanislavski calls the period of "physical embodiment." This is what Stanislavski is perhaps most famous for. Stanislavski strongly believes that it is not enough for the actors merely to represent a role (false imitation); they must live the role in the deepest way (true imitation). For Stanislavski, the actors must live the role continually; they must come to feel that they are the character they are preparing for. This they do by concentrating on the physical actions of the character, and then moving via the physical acts to the inner life of the character. This is the process of "reincarnation" discussed in Chapter 5. The Method actors are successful to the degree that they can truly live the experience of the character they are playing on stage.

The holy actors, in a similar vein, achieve their goal when they can successfully "reincarnate" the role of the paradigmatic individual. The inner life of the paradigmatic individual is reached through the discipline of imitation. This is the real core of the Rāgānugā Bhakti Sādhana. A Gauḍīya Vaiṣṇava attains the world of Vraja after total identification with the *gopī*-identity. A Cistercian regains the lost Likeness by completely "conforming to the form" presented by Christ. To achieve success, the Theravādin not only learns the Buddha's *dharma,* but physically embodies it through a life of discipline (*vinaya*). Black Elk knew he would have had to embody the role of the sacred Red Man in order to revive the Sioux nation.

Although the two systems display interesting similarities, the differences between Stanislavski's Method and the actions of the holy actors are obvious. Stanislavski designed his method of taking on the character's identity temporarily for the purpose of successful dramatic presentation. The purpose of "true acting" for the holy actor, for whom only one drama is worthy of enactment, is salvation; taking on the identity of the paradigmatic individual ontologically transforms one forever.

There is frequently another figure in this imitative structure. Often the holy actor does not imitate a transcendent model of scripture directly, but rather imitates someone physically present, or someone who was physically present in the religious community not long ago. Returning to the diagram in Figure 5, note that some individuals are positioned between the paradigmatic individual and the holy actor—the "saints." The saint is an accomplished holy actor who sometimes functions as an additional exemplary model of perfection for a religious community. *To the extent that the saint has realized in his or her own person the identity of the paradigmatic individual,* he or she functions as a paradigmatic individual. In the formative period of a religious community, the textual approach discussed above is of much greater importance. If no exemplary saints exist, there is no choice but to turn to scripture for a model of perfection. But after enough time has passed for accomplished actors or saints to emerge, an alternative path becomes possible. The holy actors may now pattern their behavior on the exemplary actions of the saint, which by definition would be harmonious with those of the paradigmatic individual. The words of Saint Paul come to mind: "Be imitators of me, as I am of Christ." A connecting chain emerges: as the saint follows the paradigmatic individual, so the holy actor follows the saint.

Rūpa's theoretical works were written in the early years of the Gauḍīya Vaiṣṇava movement. Therefore, his primary models for the Rāgānugā Bhakti Sādhana are the paradigmatic individuals of scripture, the Vrajaloka. Viśvanātha Cakravartin, however, writing over a hundred years later when the Gauḍīya Vaiṣṇava community was more firmly established, instructs the practitioner also to imitate the Vṛndāvana Gosvāmins and one's own *guru*s, all of whom the tradition declares to be saints (*siddha*s). The Gauḍīya Vaiṣṇavas have two distinct terms to designate two types of paradigmatic figures. The original Vrajaloka are the *nitya-siddha*s, the eternally perfected ones. These are the figures I have been calling the "paradigmatic individuals." Besides these paradigmatic figures there are also the *sādhana-siddha*s, the "saints" who have achieved perfection by means of spiritual practice. The Vṛndāvana Gosvāmins and the true *guru*s typically belong to this latter category.[18] The saint is thus a valuable resource for any religious community; the saint provides a model that is physically present. The Gauḍīya Vaiṣṇavas expend much energy on preserving and narrating the biographies (more technically, the essence of their deeds, *caritāmṛta*) of the saints.

The saint also features prominently in other traditions. The saint plays an important exemplary role in Cistercianism, and the Theravāda saint, the *arhat,* who differs from Buddha only by the fact that Buddha was the Pathfinder, serves as an appropriate model for achieving *nirvāṇa.* When we search for the figure of the "saint," the accomplished holy actor, in the religious situation of Black Elk, we find none. Black Elk's visionary efforts ended in tragedy; the religious world he envisioned was never fully embodied by anyone. Black Elk himself sadly admits that neither he nor anyone else was able to actually assume the role of the sacred Red Man and restore the withering tree of the Sioux nation.

Where the saint does exist, his or her actions may appear to be identical with those of the holy actor, but the cause or motivation of those actions is different. The saint and the holy actor both act in accord with the actions set down by the paradigmatic individual, but where the actions of the holy actor are the result of discipline and imitation, the actions of the saint, who has realized the true behavior of the paradigmatic individual, are the spontaneous result of the internalized role. In Gauḍīya Vaiṣṇavism, the ideal actions of the saint (*sādhana-siddha*) and the practitioner of the Rāgānugā Bhakti Sādhana are viewed as the same, but the actions of the perfected saint are understood to be *anubhāva*s (spontaneous expressions of a true inner state), while those of the practitioner are understood to be *sādhana* (intentional acts of imitative discipline designed to achieve the true inner state). These two sides of religious action are also found in the Theravāda tradition. The *arhat* naturally expresses Buddha's *dharma,* while the ordinary *bhikṣu*'s actions are molded to that *dharma* by means of the monastic discipline.[19]

The activities of the monk-like holy actors usually require the support of a "lay community." This is a community comprised of members who have not given up their ordinary social role in pursuit of a transcendent identity, but interact with and provide assistance to those who have. The members of the lay community, we must assume, receive some kind of benefit from their interaction with the holy actors. The relationship that exists between these two groups is very complex, but one feature that comes to light through this study is that the lay community functions as a supportive audience, benefiting from the dramatic actions of the holy actors. To the degree that the imitative actions of the holy actors are true, they too embody the role of the paradigmatic individual, thus making the paradigmatic individual and his or her mythical reality vividly present for the lay community. The "lay" residents of the Vraja region willingly support the serious Rāgānugā practitioners, who represent for those residents the true *gopī-bhāva.* The Theravāda *bhikṣu* is supported by the lay Buddhists; in turn the *bhikṣu* makes himself physically available to the lay community and represents Buddha's *dharma* in word and deed. The imitative ritual activity of Black Elk enables the rest of the Sioux community to have a glimpse of the world of his vision, in which the Sioux nation exists fully renewed.

Lay communities often develop alternative paths of salvation that allow much more direct access to the ultimate world of the paradigmatic individual; these paths usually involve a greater emphasis on divine grace. The Gauḍīya Vaiṣṇavas gradually develop a significant path of unmerited grace, Luther rejects the mediating role of the monks for one of a direct personal relationship of faith, and the lay Theravādins develop the path of *stūpa* worship. Nonetheless, the structure of imitative religious action remains important in a number of traditions. The holy actor "lives a myth" by imitating and thereby internalizing the role presented by a paradigmatic individual.

We began this study with the methodological aid of theoretical works concerned with the forces which maintain a dominant social reality. To be sure, religion frequently serves this end. Throughout this study, however, I have tried to go beyond these theoretical works and demonstrate that religion also, and perhaps more importantly, challenges the social definition of reality. Religion often provides a pathway of *freedom from* the socio-historical determinism of social reality. It does this by proposing an alternative transcendent identity and the means to attain it. The *bhakta*s insist that religion is not something one is born into, but something into which one must be *reborn.* The social identity, acquired by imitating a parent, is based on pragmatic tradition, but the holy actors believe that the religious identity they pursue is founded on a divine model that transcends mere pragmatism. The quest of the holy actors is a search for freedom and meaning in a transcendent reality beyond the social reality of their own limited time and space. Thus "living a myth" may be the ultimate act of freedom.

Appendix A

Translations from *Bhaktirasāmṛtasindhu*[1]

Eastern Division: Second Wave (The Method of Realization)

270. Rāgānugā is now defined:

Rāgānugā is that (method of *bhakti*) which imitates the Rāgātmikā (*bhakti*) clearly manifest in the inhabitants of Vraja.

271. Rāgātmikā is first defined in order to distinguish it from Rāgānugā.

Passion (*rāga*), which is naturally sweet, is the highest access to the beloved (i.e., Kṛṣṇa).

272. That *bhakti* which is completely absorbed in or identical with that passion is here declared to be Rāgātmikā.

273. This (Rāgātmikā Bhakti) is of two kinds: Amorous Bhakti (*kāma-rūpa*) and Relational Bhakti (*sambandha-rūpa*).

274. Certainly, as the Seventh Canto of the *Bhāgavata* (7.1.29–30) says:

> Many have fixed their minds on the Lord by means of *bhakti* that is motivated by passion, hatred, fear, or affection and have given up sin; they have attained the goal.

275. The *gopīs* through love, Kaṃsa through fear, King Śiśupala, and the other princes of Cedi through hatred, and the Vṛṣṇis through kinship, you (Pāṇḍavas) through affection, and we (Nārada and other sages) through *bhakti,* O Mighty One.

276. Because they are contrary to the favorable nature (*anukūlya*) of *bhakti,* hatred and fear are ruled out. Affection expresses friendship and thus belongs in Vaidhī Bhakti.

[1]I translate from Rūpa Gosvāmin, *Bhaktirasāmṛtasindhu,* ed. Haridāsa Dāsa (Navadvīpa: Haribol Kuṭīr, 1945), pp. 81–94.

277. Moreover, because affection can also denote love (*prema,* a stage beyond practice), it is not included here in the discussion of religious practices. "We through *bhakti*" clearly referred to Vaidhī Bhakti.

278. The statement that the goal of his (Kṛṣṇa's) enemies and friends is the same means that Brahma resembles Kṛṣṇa as a sunray resembles the sun.

279. Usually the enemies of Hari are absorbed only in Brahma. Some, who obtain the semblance of the same form (as Brahma), become immersed in the happiness of that state (of liberation).

280. As the *Brahmāṇḍa Purāṇa* says:

> Beyond the darkness is the world of the realized ones where dwell the demons killed by Hari and those who are immersed in the bliss of Brahma.

281. Those dear to Kṛṣṇa worship him by means of an attachment to some kind of passion (*rāga*), and thereby attain the nectar of his lotus-feet which consist of love.

282. Indeed, as the Tenth Canto of the *Bhāgavata* (10.87.23) says:

> The sages, whose firm minds, eyes, and secret breaths yoke them to *yoga,* contemplate (Kṛṣṇa) in their hearts; so his enemies approach him through their preoccupation. (That is, the enemies are yoked to Kṛṣṇa in a contemplation motivated by hatred.) The *gopī*s, whose thoughts are fastened upon (Kṛṣṇa's) club-like arms which are like the King of Snakes, similarly approach the nectar of his lotus-feet, and so do we who are like the *gopī*s.

283. Amorous Bhakti:

Amorous Bhakti is that (type of Rāgātmikā Bhakti) which leads the thirst for sexual enjoyment to its perfect state, since it is undertaken exclusively for the pleasure of Kṛṣṇa alone.

284. It is perfectly accomplished and brilliantly displayed in the *gopī*s of Vraja. Their particular perfect love (*prema*) attains a special sweetness. Because it is connected with the various divine love sports, the wise call it amorous (*kāma*).

285. As the Tantra says:

> Only the perfect love of the *gopī*s is celebrated as amorous.

286. Thus even Uddhava and the other male friends of the Lord wish for it.

287. But the immature love (*rati*) found in Kubjā is understood to possess an excess of common amorousness.

288. Relational Bhakti:

Relational Bhakti is that (type of Rāgātmikā Bhakti) which involves the identification of oneself as a relation—father, and so forth—to Govinda.

Here the Vṛṣṇi cowherds are considered to be the exemplary representatives (of Relational Bhakti) due to the excellence of their passion which is yet conditioned by the awareness of the divine majesty (of Kṛṣṇa).

289. The true nature of Amorous and Relational Bhakti is essentially perfect love (*prema*), and as these two are located in the eternally perfected ones (of Vraja), they are not discussed here in detail.

290. Following the twofold nature of Rāgātmikā Bhakti, Rāgānugā Bhakti is declared to be of two kinds: Imitation of Amorous Bhakti (*kāmānugā*) and Imitation of Relational Bhakti (*sambandhānugā*).

291. Those eligible for Rāgānugā Bhakti:

Those who are eligible for (Rāgānugā Bhakti) should be desirous of the attainment of the emotional state (*bhāva*) of those residents of Vraja situated exclusively in Rāgātmikā.

292. The sign of the birth of this intense desire for those emotional states is that when hearing of the sweetness of their various emotional states the mind is not dependent upon scriptural commands nor reasoning.

293. But the one engaged in Vaidhī Bhakti should rely on the commands of scripture and favorable reasoning until the time of the manifestation of one of these emotional states (*bhāva*s).

294. (The Rāgānugā practitioner) should dwell continually in Vraja, absorbed in the stories (of the cosmic drama of Vraja), remembering (*smaraṇa*) Kṛṣṇa and his beloved intimates one is most attracted to.

295. The one desirous of attaining one of the emotional states (of the Vrajaloka) should do performative acts of service (*sevā*) in a manner which imitates the Vrajaloka with both the practitioner's body (*sādhaka-rūpa*) and the perfected body (*siddha-rūpa*).

296. The injunctions, listening (*śravaṇa*), praising (*kīrtana*), etc., described for Vaidhī Bhakti are also known by the wise (to be useful) here (in Rāgānugā Bhakti).

297. Imitation of Amorous Bhakti:

Imitation of Amorous Bhakti is that thirst which imitates the Amorous Bhakti (exemplified by the *gopī*s).

298. It is of two kinds: The Desire for Sexual Enjoyment (*sambhogecchāmayī*) and The Desire to Share in Their Emotions (*tattadbhāvecchātmikā*).

299. The goal of The Desire for Sexual Enjoyment is direct amorous involvement. The goal of The Desire to Share in Their Emotions is the sweetness (or vicarious enjoyment) of the various emotions of (the people of Vraja).

300. Those who are desirous of one of these emotional states, after looking at the sweetness of the beautiful image (of Kṛṣṇa) or after hearing of the various sports, have these two ways as a means of realizing it.

301. As—

> All the great old sages, living in the Daṇḍaka forest saw the beautiful Rāma and then desired to enjoy his pleasing body.

302. They all attained the state of womanhood and were born in Gokula, and having attained Hari by means of passion were released from the ocean of worldly existence.

303. He who has great amorous desire (for Kṛṣṇa), but acts only by means of the path of injunctions (*vidhi-mārga*), becomes a queen in the city (i.e., Dvārakā).

304. As the *Mahā-Kūrma Purāna* says:

> The great-souled sons of Agni attained womanhood by means of asceticism and thereby attained Vāsudeva who is the unborn and all-pervading preserver and origin of the universe.

305. Imitation of Relational Bhakti:

Imitation of Relational Bhakti is declared by the sages to be that *bhakti* which consists of meditating on a relationship (with Kṛṣṇa)—fatherhood, and so forth—and imposing (*āropaṇa*) such a relationship on one's own self.

306. This *bhakti* is to be practiced by those practitioners desirous of parenthood, friendship, and so forth, by means of the emotions, actions, and postures of the king of Vraja (Nanda), Subala, and other appropriate models.

307. It is written in the scriptures that a certain wise old man who lived in Kurupurī became perfected worshipping, by Nārada's instruction, the image of the son of Nanda (i.e., Kṛṣṇa) as his own son.

308. Thus the *Nārāyana–vyūha–stava* says:

> Obesance to those zealous ones who are constantly meditating on Hari as a husband, a son, a companion, a brother, a father, or a friend.

309. Rāgānugā, which is the cause of the attainment of the compassion of Kṛṣṇa and his *bhakta*s, is called Puṣṭi Mārga by some (i.e., by the Vallabha Sampradāya).

Appendix B

Aṣṭa-Kālīya-Līlā-Smaraṇa-Maṅgala-Stotraṃ[1]

1. I praise Kṛṣṇa's eternal activities in Vraja in order to explain now the mental worship to be performed by those travelling on the path of passion (i.e., Rāgānugā). This mental worship achieves the service of love at the lotus-feet of the dear friend of Śrī Rādhā (i.e., Kṛṣṇa), a service which is attained by those absorbed in the activities of Vraja with eager desire, but is inaccessible to Keśa, Śeṣa, and Ādi (Brahmā, Śiva, and Ananta-Śeṣa i.e., those following the path of "liberation" [*mukti*]).
2. May we be protected by Kṛṣṇa, who at night's end leaves the bower and returns to the cowherd village, in the morning and at sunset milks the cows and eats his meals, at midday roams about playing with his friends and tending cattle, in the afternoon returns to the cowherd village, in the late evening amuses his dear ones, and at night makes love in the forest with Rādhā.
3. At night's end, I remember Rādhā and Kṛṣṇa who are awakened by the songs of parrots and cuckoo birds, both pleasing and displeasing, and by many other noises sent by a concerned Vṛndā (goddess of the forest). Arising from their bed of joy, these two are looked upon and pleased by their female friends (*sakhīs*), and though filled with desire and trembling from the passion that arises at that time, they return to the beds of their own homes, fearful of the crowing cock.
4. In the morning, I take refuge with Rādhā who, bathed and decorated, is summoned with her friends by Yaśodā to his house (Kṛṣṇa's house at Nandagrāma) where she cooks the prescribed food and enjoys Kṛṣṇa's remnants. And I take refuge with Kṛṣṇa who awakens and goes to the cow shed to milk the cows; he then is well-bathed and fed in the company of his friends.

[1]I translate Rūpa Gosvāmin, *Aṣṭa-kālīya-līlā-smaraṇa-maṅgala-stotraṃ,* Sanskrit text in De, *Vaisnava Faith and Movement,* pp. 673–75. I translate these verses for their meaning only; those wanting to appreciate the beauty of Rūpa's poetry must read the Sanskrit.

5. In the forenoon, I remember Kṛṣṇa who goes to the forest accompanied by his friends and cows, and is followed by the cowherds. Being desirous of possessing Rādhā, he goes to the bank of her pond (Rādhākuṇḍa) at the time of their secret rendezvous. And I remember Rādhā who having observed Kṛṣṇa leaves the house for the purpose of performing the sun-worship as instructed by an ascetic. She keeps her eye on the path for her own girl friend who had been sent to make arrangements with Kṛṣṇa.

6. At midday, I remember Rādhā and Kṛṣṇa who are full of desire and are decorated and made lovely by the various changes brought about by their mutual union. They are served by a host of attendants, are delighted by the jokes of the girl friends, Lalitā, etc., who arouse the god of Love, are trembling with passion and coyness, and are engaged in such playful activities as swinging, playing in the forest, splashing in the water, stealing the flute, making love, drinking honey wine, and worshipping the sun.

7. In the afternoon, I remember Rādhā who returns home, prepares various gifts for her lover, is bathed and beautifully dressed, and then is filled with pleasure at the sight of the lotus-face of her lover. And I remember Kṛṣṇa who is accompanied back to Vraja by his friends and the herd of cattle, is pleased by the sight of Śrī Rādhā, is greeted by the face of his father, and is bathed and dressed by his mother.

8. At sunset, I remember Rādhā who by means of a girl friend sends many kinds of food which she prepared for her lover, and whose heart is delighted upon eating the remnants brought back by her friend. And I remember the Moon of Vraja (Kṛṣṇa) who is well-bathed, beautifully dressed, and caressed by his mother. He goes to the cow shed and milks the cows, then returns to his house and enjoys his meal.

9. In the late evening, I remember Rādhā, who is dressed appropriately for either a light or dark night and accompanied by her group of girl friends, and who by means of a female messenger makes plans, according to Vṛndā's instructions, to rendezvous at a bower of trees of desire located on the bank of the Yamunā. And I remember Kṛṣṇa who, after watching the performance of skillful arts with the assembly of cowherds, is carefully taken home and put to bed by his affectionate mother. Later, he secretly arrives at the bower.

10. At night, I remember Rādhā and Kṛṣṇa who, being full of desire, have possessed one another. They are worshipped by Vṛndā and the many attendants, and they play with their dear friends with songs, jokes, riddles, and sweet speech, which are all associated with the circle and love dances. The minds of these two are on love and they drink prepared honey wine. Masters of love, their hearts expand by the acts performed in the bower, and they experience the various *rasa*s of love.

11. These two are delighted by the company of their girl friends and are served out of love by means of betel-nut, fragrant garlands, fans, cold water, and foot massages. After their girl friends have fallen asleep, they too drift off to sleep on their bed of flowers, murmuring the utterances of lovers full of the *rasa* of a secret love.

Glossary

Abhinavagupta Eleventh-century aestheticism from Kashmir. Abhinavagupta is the author of the *Abhinavabhāratī,* the only surviving commentary on Bharata's *Nāṭya Śāstra,* and is famous for his comparison of aesthetic experience and mystical experience.

anubhāva The aesthetic component which allows an emotion to be sensed. The *anubhāva*s make up the action of the play; these include the words and gestures of the actors.

āśraya The "vessel" of an emotion. *Āśraya* frequently refers to one of the characters of the Vraja-līlā who have an intense love for Kṛṣṇa.

bābā Literally "father." The term is used for the renunciant of the Gauḍīya Vaiṣṇava tradition who has given up ordinary life to pursue the spiritual.

Bhagavad Gītā The "Song of the Lord"; a classic Hindu text which forms part of the epic *Mahābhārata.*

Bhagavān A common name for Kṛṣṇa as personal Lord.

Bhāgavata Purāṇa A ninth- or tenth-century text that depicts the life of Kṛṣṇa; the greatest scriptural authority for the Gauḍīya Vaiṣṇavas.

bhakta A devotee, literally one who "shares" in God.

bhakti Typically translated as "devotion." I hope this book helps to expand our understanding of the complexities of *bhakti.*

Bhaktirasāmṛtasindhu A sixteenth-century text written by Rūpa Gosvāmin. In this text Rūpa presents religion in terms of Bharata's *rasa*-theory; this text is the primary sourcebook for the Rāgānugā Bhakti Sādhana.

Bharata The legendary author of the Nāṭya Śāstra.

bhāva "Emotion" or "feeling," the basis of *rasa;* frequently used to denote one of the types of emotional relationships possible with Kṛṣṇa.

Caitanya A Bengali saint (C.E. 1486–1533); the inspirational leader of Gauḍīya Vaiṣṇavism.

Caitanya-caritāmṛta The most authoritative biography of Caitanya. It was written at the beginning of the seventeenth century by Kṛṣṇadāsa Kavirāja.

dīkṣā-guru The spiritual preceptor who formally initiates the aspirant into a practice. This figure is also known as the *mantra-guru* because the initiation is by means of a *mantra.*

Gauḍīya Vaiṣṇavism A religious movement. It has its roots in sixteenth-century Bengal, was inspired by the saint Caitanya, systematized by the Vṛndāvana Gosvāmins, and is devoted to the worship of Rādhā-Kṛṣṇa.

gopī A cowherd lover of Kṛṣṇa; one of the women of Vraja.

guṇa A "constituent" of the material world; can also mean a "quality" of someone or something.

jīva The individual soul.

Jīva Gosvāmin One of the Vṛndāvana Gosvāmins who wrote numerous important philosophical texts; a nephew of Rūpa Gosvāmin.

kīrtana A devotional style of singing that involves repetition of the names of Kṛṣṇa.

Kṛṣṇadāsa Kavirāja A star pupil of the Vṛndāvana Gosvāmins; author of the *Caitanya-caritāmṛta.*

līlā "Play," both in the sense of "fun," or "game," and in the sense of "drama."

līlā-smaraṇa A meditation technique that involves visualizing the love-play of Kṛṣṇa.

mañjarī One of the female servant-companions of Rādhā.

Mañjarī Sādhana A spiritual practice that involves mentally transforming oneself into a *mañjarī* and worshipping the divine couple Rādhā-Kṛṣṇa through various acts of service.

mantra A mystical verse or sacred formula.

māyā "Illusion," that force by which Kṛṣṇa conceals or distorts his true nature.

mokṣa "Liberation," the Advaitin goal of union that is devalued by the *bhakti* theorists. Also known as *mukti.*

mukti See *mokṣa.*

mūrti A "form" of God; name for the image worshipped by Hindus.

Nanda Kṛṣṇa's adoptive father; he is the chief of the cowherds of Vraja.

Nandagrāma Nanda's village; the site of Kṛṣṇa's childhood home in Vraja.

Narottama Dāsa An important seventeenth-century Gauḍīya Vaiṣṇava figure crucially involved in the development of Mañjarī Sādhana.

Nāṭya Śāstra A compendium of the theatre containing the first textual discussion of the *rasa* theory. This text is most likely a product of the Gupta era (fourth–sixth centuries C.E.).

prema Supreme "love"; the culmination of *bhakti-rasa.*

Rādhā Kṛṣṇa's most beloved *gopī.*

rāga "Passion," specifically the passion of the Vrajaloka, the lovers of Kṛṣṇa.

rāgānugā The path that follows or imitates the passion of the Vrajaloka.

Rāgānugā Bhakti A type of devotion that follows or imitates the passion of the Vrajaloka.

Rāgātmikā Bhakti A type of perfect devotion exemplified by the Vrajaloka.

Rādhākuṇḍa Rādhā's pond; site of the midday love games of the divine couple. Rādhākuṇḍa is the residence of many serious practitioners of the Rāgānugā Bhakti Sādhana.

rasa The central focus of Indian aesthetics. *Rasa* originally meant "sap," essence," or "taste." In the context of aesthetics, it is best understood as "aesthetic enjoyment"; in the context of *bhakti* as "devotional sentiment." The Gauḍīya Vaiṣyanas equate *rasa* with divinity.

rāsa-līlā A drama depicting an episode from the life of Kṛṣṇa.

Rūpa Gosvāmin Perhaps the most influential of the Vṛndāvana Gosvāmins. Rūpa Gosvāmin is the author of the *Bhaktirasāmṛtaindhu* in which the

Rāgānugā Bhakti Sādhana and the aesthetic approach to *bhakti* are first systematically presented.

Rūpa Kavirāja A seventeenth-century figure who was censured by the Gauḍīya Vaiṣṇava community for his views on Rāgānugā Bhakti Sādhana.

sādhaka-rūpa (also *sādhaka-deha*) The "practitioner's body"; generally understood to mean the practitioner's given physical body.

sādhana Spiritual practice; means of realization.

sakhī A female friend of Rādhā.

śakti "Power"; the divine energy of Kṛṣṇa.

Sanātana Gosvāmin An important member of the Vṛndāvana Gosvāmins; the elder brother of Rūpa Gosvāmin.

sevā Typically translated as "service," though I have favored the translation "performance" or "performative acts" throughout.

siddha-rūpa (also *siddha-deha*) The "perfected body"; generally understood to mean an interior meditative body which has access to the eternal play of Kṛṣṇa.

śikṣā guru The spiritual preceptor who instructs the aspirant in a particular practice.

smaraṇa "Remembrance," a contemplative technique.

śravaṇa guru Any individual from whom the aspirant first hears about the world of Ultimate Reality.

sthāyi-bhāva A foundational or dominant emotion; the basis of a *rasa*.

svarūpa "Essential nature."

taṭastha Literally "standing on the border"; it refers to a liminal state of being.

ṭhākura ghara The "god room" which houses the images of one's favorite deities.

Ujjvalanīlamaṇi A sequel to the *Bhaktirasāmṛtasindhu* which focuses on discussion of the amorous sentiment.

vaidhi That which is related to injunctions.

Vaidhī Bhakti A type of devotion which follows the injunctions of scriptural commands.

Vaiṣṇava Relating to Viṣṇu/Kṛṣṇa; one who worships Viṣṇu/Kṛṣṇa.

vāsanā An unconscious karmic impression; an inclination; a seed for later experience.

vana-yātrā "Procession through the forest"; a pilgrimage which involves visiting the various sites associated with the life of Kṛṣṇa in Vraja.

vibhāva The aesthetic component that first excites an emotion. The *vibhāva*s are the environmental conditions of an emotion; these include both the characters and the setting of a play.

viṣaya The object of an emotion. In Gauḍīya Vaiṣṇava texts, *viṣaya* usually refers to some manifestation of Kṛṣṇa.

Viśvanātha Cakravartin An important commentator on the works of Rūpa Gosvāmin who lived in Vṛndāvana during the second half of the seventeenth century and first half of the eighteenth century.

Vraja Site of Kṛṣṇa's cosmic drama; the region associated with the playful life of Kṛṣṇa roughly corresponding to the western part of the Mathurā District of the state of Uttar Pradesh.

Vraja-līlā The love-play of Kṛṣṇa and his companions which took place in the land of Vraja.

Vrajaloka The lovers of Kṛṣṇa who reside in Vraja; these are the paradigmatic individuals for the Gauḍīya Vaiṣṇava tradition.

Vṛndāvana The forest site of Kṛṣṇa's nightly tryst with Rādhā; a pilgrimage town located on the bank of the Yamunā about eighty miles south of Delhi; the spiritual center of Gauḍīya Vaiṣṇavas in Vraja.

Vṛndāvana Gosvāmins A group of six theologians sent to Vṛndāvana by Caitanya in the sixteenth century to establish Vṛndāvana as a spiritual center and provide a systematic foundation for the emerging Gauḍīya Vaiṣṇava movement.

vyabhicāri-bhāva The aesthetic component that supports and fosters the dominant emotion. The *vyabhicāri-bhāva*s are secondary accompanying emotions and are sometimes called *sañcārī-bhāva*s.

Yaśodā The adoptive mother of Kṛṣṇa.

yoga-pīṭhāmbuja A "lotus-place of union"; a meditative diagram depicting a place where Rādhā and Kṛṣṇa come together for love.

Notes

Chapter 1

1. This religious movement is also known as Bengal Vaiṣṇavism or the Caitanya Sampradāya. The latter name is perhaps best, since this movement has followers outside the area of Bengal; however, I will retain the name Gauḍīya Vaiṣṇavism because it is the most familiar among Western scholars.

2. The term "reality" is a bit problematic. At present I would like to use the term loosely, interchanging it with the word "world." Peter Berger and Thomas Luckmann define reality as "a quality appertaining to phenomena that we recognize as having a being independent of our own volition (we cannot wish them away)" (*The Social Construction of Reality* [New York: Doubleday & Co., 1966], p. 1). However, I mean to include in my own usage of the term what Clifford Geertz calls "perspective." "A perspective is a mode of seeing, in that extended sense of 'see' in which it means 'discern,' 'apprehend,' 'understand,' or 'grasp.' It is a particular way of looking at life, a particular manner of construing the world" (*The Interpretation of Cultures* [New York: Basic Books, Inc., 1973], p. 110).

3. The anthropological implications of "world-openness" have been developed by Arnold Gehlen. See Peter Berger and Thomas Luckmann, "Arnold Gehlen and the Theory of Institutions," *Social Research* 32, no. 1 (1965): 110 ff.

4. William James was one of the first among social scientists to analyze the multiple nature of reality. See *Principles of Psychology*. 2 vols. (New York: Henry Holt & Co., 1890), 2: 283–322. See also Alfred Schutz, *Collected Papers I: The Problem of Social Reality,* ed. Maurice Natanson (The Hague: Martins Nijhoff, 1973), pp. 340–47.

5. My own answer to this question is influenced by sociologists of knowledge such as Alfred Schutz and his recent interpreters Peter Berger and Thomas Luckmann. I believe that their theory of socialization (and the important part the "role" plays in it) is a useful tool with which to approach an understanding of how one comes to inhabit a specifically "religious" reality, though I will be applying the theory to an arena untried by its creators. See Schutz, *Collected Papers,* and Berger and Luckmann, *The Social Construction of Reality*.

6. Compare this notion of institution with Geertz's notion of "symbol." Geertz, *The Interpretation of Culture,* pp. 91–94.

7. Berger and Luckmann, *Social Construction of Reality,* p. 131 ff.

8. See Theodore R. Sarbin, "Role Theory," in Gardner Lindsey, ed., *Handbook of Social Psychology* (Cambridge: Addison-Wesley, 1954), pp. 223–58.

See also Bruce J. Biddle and Erwin J. Thomas, eds., *Role Theory: Concepts and Research* (New York: John Wiley & Sons, 1966).

9. See Berger and Luckmann, *Social Construction of Reality,* p. 132.

10. Previous attempts to use role theory as an instrument for interpreting religious action would include: Victor Turner's notion of "root paradigms," as seen in his discussion of Thomas Becket in *Dramas, Fields, and Metaphors* (Ithaca: Cornell University, 1974); Hjalmar Sudén's theological application of role theory, as discussed in *Die Religion und Die Rollen: Ein Psychological der Fromigkeit,* translated by Herman Muller and Suzanne Okman (Berlin: Topelmann, 1966); and Donald Capps, "Sudén's Role-Taking Theory: The Case of John Henry Newman and His Mentors," *Journal for the Scientific Study of Religion* 21, no. 1 (March 1982): 58–70. Perhaps my favorite, however, is expressed in a novel by Nikos Kazantzakis appearing under the English title *The Greek Passion* (New York: Simon & Schuster, 1953).

11. See Mircea Eliade, *The Sacred and the Profane* (New York: Harcourt, Brace & World, 1959).

12. Joachim Wach argues that the religious experience is a response to what is experienced as the Ultimate Reality. See Joseph M. Kitagawa, ed., *The Comparative Study of Religion,* (New York: Columbia University, 1958), p. 30.

13. Schutz, *Collected Papers I,* pp. 231–32.

14. Berger and Luckmann emphasize this point; see *Social Construction of Reality,* p. 145.

15. Ralph C. Beals, "Religion and Identity," *Internationales Jahrbuch für Religionssoziolgie* 11 (1978): 147. In this brief but interesting article, Beals shows how the "transcendent identity" serves to free one from historical determinism.

16. William James writes of religion as relating to an "unseen order." See *The Variety of Religious Experience* (New York: Longman, Green, and Co., 1902), p. 58.

17. Mircea Eliade speaks of the important role exemplary mythic models play in providing a guide for all religious acts. See especially *The Myth of the Eternal Return* (Princeton: Princeton University, 1954).

18. Buddha and Christ are classical examples of "paradigmatic individuals." For previous uses of this term, see Karl Jaspers, *Socrates, Buddha, Confucius, and Jesus: The Four Paradigmatic Individuals,* translated by Ralph Manheim (New York: Harvest Books, 1957), and A. S. Cua, *Dimensions of Moral Creativity: Paradigms, Principles, and Ideals* (University Park: Pennsylvania State University, 1978).

19. See, for example, Melford E. Spiro, "Religion: Problems of Definition and Explanation," in Michael Banton, ed., *Anthropological Approaches to the Study of Religion* (London: Tavistock Publications, 1966), p. 98.

20. Mircea Eliade, *Myth and Reality* (New York: Harper & Row, 1963), p. 19. Bronislaw Malinowski also made it clear that myth is not merely a story to be told, but is a reality to be lived. See *Magic, Science and Religion* (New York: Doubleday & Co., 1948), pp. 108 ff.

21. The term "imitation" is an important one for the present study; therefore, it is necessary to establish what I mean by the term. The term has two connotations for the English speaker. First, the term is used to refer to something that is a fake copy or counterfeit. The second use of the term, used particularly by sociologists and psychologists, refers to the performance of an act which involves the copying of patterns of behavior and thought of other individuals. I use the term imitation only in this latter sense. Moreover, I follow Eliade's use of the term; for Eliade, "imitation" of a religious paradigm guarantees the authenticity of an act.

22. Yoshita S. Hakeda, *Kūkai: Major Works* (New York: Columbia University, 1972), p. 98.

23. The Penitentes of New Mexico, for example, dramatically imitate the Passion of Christ as part of their practice. See the practices outlined by Martha Weigle, *Brothers of Light, Brothers of Blood: The Penitentes of the Southwest* (Albuquerque: University of New Mexico, 1976).

24. Anthony Wallace also defines salvation as a transformation of identity. See *Religion: An Anthropological View* (New York: Random House, 1966), pp. 138–57.

25. Emory Sekaquaptewa, "Hopi Indian Ceremonies," in Walter Capps, ed., *Seeing with a Native Eye* (New York: Harper & Row, 1976), p. 39.

26. Timothy J Wiles, *The Theater Event* (Chicago: University of Chicago, 1980), p. 14 ff.

27. Many actors have attested to this experience. My favorite is Dustin Hoffman's confession that a whole new world of meaning, a whole new perspective, was opened up when he played the role of a woman in the film *Tootsie.* "I'm telling you, if you are a woman for a month, the world is a different experience in ways you never imagine. . . . My wife tells me that playing the part altered me" ("Tootsie Taught Dustin Hoffman about Sexes," by Leslie Bennetts, *New York Times,* 21 December 1982, p. C11).

28. Constantin Stanislavski, *An Actor Prepares,* translated by Elizabeth R. Hapgood (New York: Theatre Arts, 1946), p. 278. Note the striking similarity to Hopi masking.

29. This is an interesting choice of terms. As far as I am aware, Stanislavski had little knowledge of Indian religious thought. For more on Stanislavski's "reincarnation," see P. V. Simonov, "The Method of K. S. Stanislavski and the Physiology of Emotion," in Sonia Moore, ed., *Stanislavski Today* (New York: American Center for Stanislavski Theatre Art, 1973), p. 73; and Sonia Moore, *The Stanislavski System* (New York: Penguin Books, 1965), p. 73.

30. Stanislavski, *An Actor Prepares,* p. 294.

CHAPTER 2

1. Plato, *The Republic* 10, in *Great Dialogues of Plato,* translated by W.H.D. Rouse, edited by Eric H. Warmington and Philip G. Rouse (New York: The New American Library, 1956), pp. 393–422.

2. Aristotle, *Poetics* 2–6, translated by Kenneth A. Telford (Chicago: Henry Regnery Co., 1970), pp. 4–14.

3. This term will be explained in detail below.

4. Jan Gonda, gen. ed., *A History of Indian Literature,* 10 vols. (Wisebaden: Otto Harrassowitz, 1973), vol. 5: *Indian Poetics,* by Edwin Gerow, p. 245.

5. A compendium of the theatre. S. K. De assigns the legendary author of the *Nāṭya-śāstra,* Bharatamuni, to the period ranging between the second century B.C.E. to the second century C.E., while admitting the present form of the text may be as late as the eighth century (*History of Sanskrit Poetics,* 2 vols. [Calcutta: K. L. Mukhopadhyay, 1960], 1: 18). Gerow generally agrees with this historical placement of the formative period, but suggests that it should not be dated later than the sixth century. "It is thus roughly contemporaneous with the great flowering of dramatic and other literature under the patronage of the Gupta kings (fourth to sixth centuries), and it reflects the cultural and aesthetic realities of that flowering" (*Indian Poetics,* p. 245).

6. See *Ṛg Veda* 10.90 and *Bṛhadāraṇyaka Upaniṣad* 1.4.

7. *Nāṭya–śāstra* 1.114. The edition I cite is one published in Devanagari script with the commentary of Abhinavagupta, which was edited and translated into Hindi by Madhusudan Shastri (Varanasi: Banaras Hindu University, 1971).

8. Aristotle, *Poetics* 6, pp. 11–14

9. See *Sāhitya–darpaṇa* 3.29.

10. See *Nāṭya–śāstra* 7.

11. A good discussion of the components of dramatic experience is found in De, *History of Sanskrit Poetics,* vol. 2.

12. *Nāṭya–śāstra* 6.32. *Vibhāvānubhāvavyabhicāri-samyogād rasaniṣpattiḥ.*

13. M. Christopher Byrski, *Concepts of Ancient Indian Theatre* (New Delhi: Munshiram Manoharlal Publishers, 1973), pp. 150–52.

14. We know very little of Bhaṭṭa Nāyaka. His writings survive only in the works of Abhinavagupta.

15. *Abhinavabhāratī of Abhinavagupta,* translated by Raniero Gnoli in *The Aesthetic Experience According to Abhinavagupta,* rev. 2nd ed. (Varanasi: Chowkhamba, 1968), p. 45.

16. See Suresh Dhayagude, *Western and Indian Poetics: A Comparative Study* (Pune: Bhandarkar Oriental Research Institute, 1981), p. 176 ff.

17. See S. K. De, *Sanskrit Poetics as a Study of Aesthetics,* with notes by Edwin Gerow (Berkeley: University of California Press, 1963), p. 21.

18. Gnoli, *Asthetic Experience According to Abhinavagupta,* p. xxi.

19. Ibid., p. 48.

20. J. L. Masson and M. V. Patwardhan, *Śāntarasa and Abhinavagupta's Philosophy of Aesthetics* (Poona: Bhandarkar Oriental Research Institute, 1969), p. 21.

21. Ibid., p. 32.

22. K. C. Pandey, *Abhinavagupta: An Historical and Philosophical Study* (Varanasi: Chowkhamba Sanskrit Series, 1963), pp. 20–22.

23. Masson and Patwardhan, *Śāntarasa,* p. 41.

24. *Tantrāloka,* cited and translated by Gnoli, *Aesthetic Experience According to Abhinavagupta,* pp. xxxviii–xxxix.

25. J. L. Masson and M. V. Patwardhan, *Aesthetic Rapture* (Poona: Deccan College, 1970), p. 3.

26. *Locana,* cited and translated by Masson and Patwardhan, *Śāntarasa,* pp. 50–51.

27. One of the first significant developments in Bharata's *rasa* theory was its extension into poetics. This new dimension developed into a school of criticism associated with the name Ānandavardhana, a ninth-century writer who composed the *Dhvanyāloka.* The new school was called *dhvani,* which translates as "tone" or "suggestion." The *dhvani* theory draws upon the earlier theoretical discussions of the Indian philosophers of language. Ānandavardhana recognized that, besides its power to denote and indicate literal meaning, a word has the power to "suggest" an inexpressible meaning. Applying this notion to poetry, Ānandavardhana distinguished the essentials of poetry into two parts, the expressed and the unexpressed, and declared the unexpressed to be the "soul" or essence of poetry. Like the *rasa* theory of Bharata, this theory focused on the emotions. The *dhvani* theorists assumed that emotions could be named, yet this basically differs from expression. Moods and feelings, which were held to be the essence of the poem, could not be directly expressed—at best they could only be suggested. The "suggested" (*dhvani*) part of the poem was then declared by these theorists to be *rasa.*

28. Masson and Patwardhan, *Aesthetic Rapture,* p. 18.

29. Masson and Patwardhan, *Śāntarasa,* p. 89.

30. Masson and Patwardhan, *Śāntarasa,* pp. vii–viii.

31. Gnoli, *Aesthetic Experience According to Abhinavagupta,* p. xlvi.

32. Eliot Deutsch, *Advaita Vedānta* (Honolulu: University of Hawaii Press, 1973), p. 58.

33. Note here the sharp contrast with Plato's conclusions regarding the value of aesthetic experience toward the realization of truth.

34. Masson and Patwardhan, *Śāntarasa,* p. 23.

35. Mysore Hiriyanna, *Art Experience* (Mysore: Kavyalaya Publishers, 1954), p. 28.

36. *Locana,* cited and translated by Masson and Patwardhan, *Śāntarasa,* p. 55.

37. *Nāṭya–śāstra* 7.31.

38. At least he was the first mentioned by Abhinavagupta. The commentary of Abhinavagupta (*Abhinavabhāratī*) is the only commentary on Bharata's *Nāṭya-śāstra* to survive, and most of what we know of the others comes from the summaries in this single commentary.

39. My translation of the Sanskrit text of the *Abhinavabhāratī* in Gnoli, *Aesthetic Experience According to Abhinavagupta,* p. 3.

40. Y. S. Walimbe, *Abhinavagupta on Indian Aesthetics* (Delhi: Ajanta Publications, 1980), p. 16.

41. Gnoli, *Aesthetic Experience According to Abhinavagupta,* p. xviii.
42. Masson and Patwardhan, *Aesthetic Rapture,* p. 35.
43. A. Berriedale Keith, *The Sanskrit Drama* (London: Oxford University Press, 1924), p. 321.
44. Masson and Patwardhan, *Aesthetic Rapture,* p. 33.
45. A major study of this work was completed by V. Raghavan, *Bhoja's Śṛṅgāra Prakāśa,* rev. 3rd ed. (Madras: Punarvasu, 1978).
46. Ibid., p. 423.
47. Ibid., p. 424. See the Sanskrit in Devanagari cited by Raghavan.
48. The edition of the *Sāhitya–darpaṇa* I cite is one published in Devanagari script, which was edited and translated into Hindi by Salagrama Sastri (Delhi: Motilal Banarsidass, 1977).
49. *Sāhitya–darpaṇa* 3.18.
50. Ibid., 3.19.
51. My translation of the Sanskrit text of the *Abhinavabhāratī* in Gnoli, *Aesthetic Experience According to Abhinavagupta,* p. 3.
52. Śaṅkuka is also known only through Abhinavagupta's *Abhinavabhāratī.*
53. *Abhinavabhāratī.* See Gnoli, *Aesthetic Experience According to Abhinavagupta,* p. 78.
54. See Raghavan, *Bhoja's Śṛṅgāra Prakāśa,* p. 475.
55. Sivaprasad Bhattacharya, "Bhoja's Rasa-Ideology and Its Influence on Bengal Rasa-Śāstra," *Journal of the Oriental Institute* (University of Baroda) 13, no. 2 (December 1963): 107. I am grateful to Robert D. Evans for providing me with this article.
56. *Sāhitya–darpaṇa* 3.1.
57. See Karl H. Potter, *Presuppositions of India's Philosophies* (Westport: Greenwood Press, 1963), pp. 150–85. Potter's discussions of "progress" (*jātivāda*) and "leap" (*ajātivāda*) philosophies are also relevant.
58. For a detailed discussion of this issue, see V. Raghavan, *The Number of Rasa-s* (Madras: The Adhar Library and Research Centre, 1967).
59. Masson and Patwardhan, *Śāntarasa,* pp. 130–31. This is a translation from the *Abhinavabhāratī.* I can do no better than Masson and Patwardhan in translating this difficult passage.
60. This is the main argument of Masson and Patwardhan in their study *Śāntarasa.* Edwin Gerow and Ashok Aklujkar, however, argue that, since for Abhinava all *rasa*s resolve into a common experience of repose (*viśrānti*), a separate *rasa* to designate this common experience is unnecessary and even awkward for Abhinava to fit into his system. See Edwin Gerow and Asok Aklujkar, "On Śānta Rasa in Sanskrit Poetics," *Journal of the American Oriental Society* 11, no. 1 (January–March 1972): 80–87.
61. Raghavan, *Bhoja's Śṛṅgāra Prakāśa,* pp. 452–53.
62. Ibid., p. 453.

CHAPTER 3

1. Vṛndāvana is a town located in Vraja. Vraja is the name used to designate the land in which the mythical life of Kṛṣṇa the cowherd takes place. It is important to note that the names Vṛndāvana and Vraja refer both to mythical places and to regions of India about eighty miles south of Delhi, which were settled by the Six Gosvāmins and are today the focus of much Vaiṣṇava pilgrimage activity.

2. S. K. De, *The Early History of the Vaiṣṇava Faith and Movement in Bengal* (Calcutta: Firma K. L. Mukhopadhyay, 1961), p. 118.

3. Masson and Patwardhan, *Aesthetic Rapture,* p. 4.

4. This text is the major sourcebook for a study of the Rāgānugā Bhakti Sādhana. Its contribution to the establishment of religious aestheticism in the mode of *bhakti* cannot be overestimated. The edition of the *Bhaktirasāmṛtasindhu* I cite is one published in Bengali script, which was edited by (and includes a Bengali translation by) Haridāsa Dāsa (Navadvīpa: Haribol Kutir, 1945). I believe this is the most authoritative edition of the text available. It includes the valuable commentaries of Jīva Gosvāmin, Viśvanātha Cakravartin, and Mukundadāsa Gosvāmin. I read portions of the *Bhaktirasāmṛtasindhu* in Chicago with Prof. Edwin Gerow and in Vṛndāvana with Dr. Acyuta Lal Bhatta. I am grateful to both for their time and insightful comments.

5. *Abhinavabhāratī,* Śānta-prakaraṇa. See Masson and Patwardhan, *Śānta Rasa,* p. 139.

6. Vopadeva was a thirteenth-century Marathi, claimed by some to be the author of the *Bhāgavata Purāṇa,* though this claim is most probably false (see J. N. Farquhar, *An Outline of the Religious Literature of India* [Delhi: Motilal Banarsidass, 1920], pp. 231, 234). The *Muktāphala* was published with the commentary of Hemādri, by Calcutta Oriental Series, Isvara Chandra Sastri and Haridasa Vidyabagisa, eds. (Calcutta: Badiya Nath Dutt, 1920), no. 5.

7. Ibid., p. 187.

8. Rūpa indicates a knowledge of this text (*Ujjvalanīlamaṇi* 15.151).

9. It remains so today. See, for example, Swami Sivananda, *Essence of Bhakti Yoga* (Shivanandanagar: Divine Life Society, 1981). This work is shot through with the *rasa* theory of Rūpa Gosvāmin, though it fails to even mention his name.

10. *Jīva Gosvāmin, Bhagavat Sandarbha,* text in Devanagari script, Chinmayi Chatterjee, ed. (Calcutta: Jadavpur University, 1972), pp. 22 ff.

11. *Rasa* is best translated in the Upaniṣads as "sap" or "essence."

12. The *Śānta-bhakti-rasa* is discussed in the *Bhaktirasāmṛtasindhu* 3.1. The four main subsequent *bhakti-rasa*s are discussed in the remaining sections of the third part.

13. Edwin Gerow, first draft of essay, "Rasa as a Category of Literary Criticism," in Rachel Van M. Baumer and James R. Brandon, eds., *Sanskrit Drama in Performance* (Honolulu: University Press of Hawaii, 1981). Unfortunately, this quotation was edited out of the final copy.

14. De, *Vaiṣṇava Faith and Movement,* p. 177. Emphasis mine.

15. See A. K. Ramanujan, *Speaking of Śiva* (Baltimore: Penguin Books, 1973). In the introduction to this work, Ramanujan argues that *bhakti,* with its emphasis on "action," is a reversal of much of the traditional "passive" Hindu norms.

16. See Sonia Moore, *The Stanislavski System* (*Digested from the Teachings of Konstantin S. Stanislavski*) (Middlesex: Penguin Books, 1965), p. 5.

17. Recent scholars have begun to notice the influence of Bhoja on Gauḍīya Vaiṣṇava aesthetics. See Sivaprasad Bhattacarya, "Bhoja's Rasa-Ideology and Its Influence on Bengal Rasa-Śāstra," in *Journal of the Oriental Institute* (Baroda: University of Baroda) 13, no. 2 (December 1963): 106–19; and S. N. Goshal, *Studies in Divine Aesthetics* (Santiniketan: Visva-Bharati, 1974), pp. 44–45.

18. Compare this to Stanislavski's notion of the actor's "re-incarnation." See above, p. 38.

19. Ananda Coomaraswamy, "Līlā," *Journal of the American Oriental Society* 61 (1941): 91.

20. Edwin Gerow, "The *Rasa* Theory of Abhinavagupta and Its Application," in Edward C. Dimock, Jr., et al., eds., *Literatures of India* (Chicago: University of Chicago, 1974), p. 227.

21. Rūpa suggests that this *sthāyi-bhāva* of Kṛṣṇa-rati may be dependent on a special *bhakti-vāsanā* (BRS 2.1.6).

22. Again Rūpa seems particularly close to Bhaṭṭa Lollaṭa. Also, if the whole world becomes drama, then the distinction between the *sthāyi-bhāva* (world) and *rasa* (drama) dissolves.

23. Rabindranath Tagore translated the Sanskrit word *sādhana* as "realization" (see *Sādhana* [Madras: Macmillan, 1913]). *Sādhana* is that means by which an ideal religious world is realized and is therefore often translated as "means of realization" or, simply, "religious practice."

CHAPTER 4

1. *Caitanya–caritāmṛta,* cited by Edward C. Dimock, Jr., "Hinduism and Islam in Medieval Bengal," in *Aspects of Bengali History and Society,* Rachel Van M. Baumer, ed. (Honolulu: University Press of Hawaii, 1975), p. 6.

2. Ibid., p. 6.

3. Norvin Hein, in an examination of the development of the worship of Rādhā, supports the notion that the erotic aspects of the Kṛṣṇa cult gained in importance under Muslim rule. See "Rādhā and Erotic Community" (especially pp. 122–124) in John Stratton Hawley and Donna Marie Wulff, eds., *The Divine Consort: Rādhā and the Goddesses of India* (Berkeley: Graduate Theological Union, 1982), pp. 116–24. Elsewhere, however, Hein argues for a much earlier shift away from Kṛṣṇa's heroic side. He maintains that the erotic cult was a response of those "living in the strait-jacket of orthodox Hinduism" during the

"extraordinary restraint" of the Gupta Age, which brought upon the population a widespread elimination of options, a narrowing of alternatives, a subjection of life to unyielding requirements, and the beginning of a sense of bondage" ("A Revolution in Kṛṣṇaism: The Cult of Gopāla," *History of Religions* 25, no. 4 (May 1986): 309–10). Others have presented this development in psychological terms. See J. L. Masson, "The Childhood of Kṛṣṇa: Some Psychoanalytic Observations," *Journal of the American Oriental Society* 94, no. 4 (1974): 454–59; and Robert Goldman, "A City of the Heart: Epic Mathurā and the Indian Imagination," *Journal of the American Oriental Society* 106, no. 3 (1986): 471–83.

4. W. G. Archer, *The Loves of Krishna in Indian Painting and Poetry* (New York: Grove Press, 1957), p. 72.

5. Ibid., p. 73.

6. David R. Kinsley, *The Sword and the Flute* (Berkeley: University of California Press, 1975), p. 62.

7. Joseph T. O'Connell, "Social Implications of the Gauḍīya Vaiṣṇava Movement" (Ph.D. dissertation, Harvard University, 1970), p. 206. Hein speaks of "the diversion of the ancient Vaiṣṇava tradition's public hope into this unassailable private world" (*The Divine Consort,* p. 123).

8. O'Connell highlights the fact that a great many of the early Gauḍīya Vaiṣṇava leaders were in some way connected to the Muslim court at Gauḍa (ibid., pp. 351–362).

9. Narahari Cakravarti, *Bhaktiratnākara,* Navīnakṛṣṇa Paravidyālaṃkāra, ed. (Calcutta: Gauḍīya Maṭh, 1940), p. 28. This is a seventeenth-century Bengali text which gives the history of the Gauḍīya Vaiṣṇava movement to that date.

10. R. C. Majumdar, *History of Mediaeval Bengal* (Calcutta: G. Bharadwaj & Co., 1973), p. 53.

11. We read in the *Bhaktiratnākara* that Rūpa and Sanātana supported and tried to be involved in traditional Hindu cultural activities while functioning as court ministers.

12. O'Connell, "Social Implications," p. 174.

13. Ibid., p. 171.

14. Beals, "Religion and Identity," p. 147.

15. Quoted in Dinesh Chandra Sen, *Chaitanya and His Age* (Calcutta: University of Calcutta, 1924), p. 220.

16. A. K. Ramanujan, *Speaking of Śiva* (Baltimore: Penguin Books, 1973), p. 27.

17. The highest *līlā* is often called the Vṛndāvana-līlā by Western scholars of Gauḍīya Vaiṣṇavism. The term Vraja-līlā, however, is more inclusive and therefore more accurate, since many of the important *līlā*s take place in parts of Vraja other than Vṛndāvana.

18. See Ananda K. Coomaraswamy, "Līlā," *Journal of the American Oriental Society* 61 (1941): 98–101.

19. Edward C. Dimock, Jr., *Place of the Hidden Moon* (Chicago: University of Chicago Press, 1966), p. 194.

20. De, *Vaiṣṇava Faith and Movement,* p. 223.

21. Revealed scripture, or *śabda,* is considered to be the most authoritative source of knowledge by the Gauḍīya Vaiṣṇava tradition; and the *Bhāgavata-Purāṇa* is considered to be the most complete scripture.

22. Those not familiar with the Kṛṣṇa-līlā are referred to David R. Kinsley, *The Divine Player: A Study of Kṛṣṇa Līlā* (Delhi: Motilal Banarsidass, 1979), and W. G. Archer's delightful study, *The Loves of Krishna in Indian Painting and Poetry* (New York: Grove Press, 1957).

23. "He is like an actor (*naṭa*) who assumes and then gives up the form of a fish, etc." *(Bhāgavata Purāṇa* 1.15.35).

24. Bhagavad-gītā 11.3.

25. See Rudolf Otto, *The Idea of the Holy,* translated by John W. Harvey (London: Oxford University Press, 1923), pp. 12–24 and 31–40.

26. *Bhāgavata Purāṇa* 10.8.44.

27. The edition of the *Bṛhad-bhāgavatāmṛta* I have used is one recently published in Devanagari script, which was edited and translated into Hindi by Śyāma Dāsa (Vrindaban: Harinam Press, 1975). An English translation of this text does exist (*Sri Brihat Bhagavatamritam,* translated by Bhakti Prajnan Yati [Madras: Gaudiya Math, 1975]), though, with all due respect to the translator, it is unreliable and often misleading.

28. Shrivatsa Gosvamin, personal conversations in Vṛndāvana, 1981–82.

29. One might wonder whether Sanātana is suggesting here that the Hindu ideals could still be realized on the sociopolitical level in southern India, which is beyond the control of the Muslims.

30. *Bṛhad-bhāgavatāmṛta* 1.7.141.

31. Ibid., 1.7.154–56.

32. *Bhagavat Sandarbha* (Jadavpur University edition), pp. 1–2.

33. Shrivatsa Gosvamin of Vṛndāvana explains that one will never understand Hinduism until one realizes that for the Hindus there are 600 million gods (the Hindu population of India) *and* that God is one.

34. The models (or *āśrayas*) are located within the discussion of the *bhāvas* and *rasas* of *bhakti.*

35. Rūpa's placement of the *śānta-rasa* in his religious schema is very much like Rāmānuja's placement of Śaṅkara's *mokṣa* in his. The *śānta-rasa* and *mokṣa* are both reduced to penultimate goals.

36. These three emotional states are discussed in *Bhaktirasāmṛtasindhu,* chapters 3.2, 3.3, and 3.4

37. It is interesting to note that in this tradition a son of God is not a very high position.

38. The edition of the *Ujjvalanīlamaṇi* I cite is one published in Bengali script, which was edited by Puridāsa (Vṛndāvana: Haridāsa Śarmana, 1954). This edition includes the commentaries of Jīva Gosvāmin and Viśvanātha Cakravartin.

39. For a discussion of the five kinds of female friends, a category of characters which was later to assume major importance, see *Ujjvalanīlamaṇi* 4.46–55.

A dual object of love is suggested by these *sakhīs*. This will become increasingly important for the development of Mañjarī Sādhana. See the fourth section of Chapter 6.

40. See, for example, *Ujjvalanīlamaṇi* 4.5. Rūpa declares Rādhā to be the *hlādinī-śakti*, an aspect of the *svarūpa-śakti*.

41. Jīva Gosvāmin, *Bhagavat–Sandarbha*, p. 32.

42. Ibid., p. 32.

43. Ibid., p. 33. See also *Viṣṇu Purāṇa* 6.7.61.

44. Ibid., p. 32.

45. Jīva Gosvāmin, *Paramātma Sandarbha*, text in Devanagari script, Chinmayi Chatterjee, ed. (Calcutta: Jadavpur University, 1972), p. 38.

46. Jīva Gosvāmin, *Paramātma Sandarbha*, pp. 37 ff.

47. Sudhindra C. Chakravarti, *Philosophical Foundation of Bengal Vaisnavism* (Calcutta: Academic Publishers, 1969), p. 47.

48. Ibid., p. 93.

49. Jīva Gosvāmin, *Bhagavat Sandarbha*, p. 101.

50. For a definition and further discussion of this perfect form of *bhakti*, see the second section of Chapter 5.

CHAPTER 5

1. Rudolf Otto, *Christianity and the Indian Religion of Grace* (Madras: Christian Literature Society for India, 1929).

2. Ibid., p. 11.

3. We must also pause to consider Otto's particular understanding of Christianity. It is significant to note that both Otto and Söderblom were early twentieth-century Lutherans.

4. Otto, *Christianity and Indian Religion*, p. 8. Emphasis mine.

5. Ibid., p. 7.

6. Ibid., p. 41. Emphasis mine.

7. Nathan Söderblom, *The Living God: Basal Forms of Personal Religion* (London: Oxford University Press, 1931).

8. Ibid., p. 133.

9. Ibid., pp. 104–5. Emphasis mine.

10. Ibid., pp. 62 and 134.

11. I find support for my affimation in the works of the Buddhist scholar Stephan Beyer. Writing on *bhakti* in the *Bhagavad-gītā* he comments: "It is clear that *yoga* is integral to the practice of *bhakti:* it is not the rather vague emotional dependence and devotionalism denoted by the term in current usage, but rather a specific contemplative activity, the iconographic visualization of the god—precisely the meditative technique that forms the episodic core of the Buddhist vision quest" ("Notes on the Vision Quest in Early Mahāyāna," in *Prajñāpāramitā and Related Systems: Studies in Honor of Edward Conze,* Lewis Lancaster and Luis Gomez, eds. [Berkeley: Berkeley Buddhist Studies Series, 1977], p. 333).

12. See Rāmānuja, *Brahma-Sūtra Śrī Bhāṣya* 1.1.

13. Certain schools of *bhakti,* of course, do place heavy, even exclusive, emphasis on unmerited grace. This is the position of the Southern (Teṅgalai) or Cathold (Mārjāra-nyāya) School of Śrī Vaiṣṇavism, the probable source of Otto and Söderblom's ideas of *bhakti.*

14. Jīva Gosvāmin, for example, defines *bhakti* as having a twofold nature: means (*sādhana*) and end (*sādhya*). See his commentary, BRS 1.2.1.

15. Mukundadāsa Gosvāmin spells this out clearly in his commentary; see BRS 1.2.6.

16. BRS 1.2.147, 172, 175, 181.

17. See John S. Hawley, *At Play with Krishna: Pilgrimage Dramas from Brindavan* (Princeton: Princeton University Press, 1981); and Norvin Hein, *The Miracle Plays of Mathurā* (New Haven: Yale University Press, 1972).

18. See Berger and Luckmann, *Social Construction of Reality,* p. 75.

19. *Caitanya-caritāmṛta* 1.3.14, cited by Dimock, *Place of the Hidden Moon,* p. 193.

20. *Caitanya-caritāmṛta* 2.8.223–25 (Gauḍīya Maṭh Edition). This text quotes *Bhāgavata Purāṇa* 10.87.23.

21. Stanislavski, *An Actor Prepares,* pp. 14–15.

22. Simonov, "The Method of K. S. Stanislavski," p. 41.

23. Swami Sivananda, *Practical Lessons in Yoga* (Rishikesh: Divine Life Society, 1978), p. 70.

24. Moore, *The Stanislavski System,* p. 22.

25. Stanislavski, *An Actor Prepares,* p. 13.

26. It must be remembered that an actor in the Russian theatre rehearsed a part for many months and strove to "live the part" in hundreds of rehearsals, workshops, and performances before the play was formally staged. Much of Stanislavski's theory pertains to the actor's efforts to get in touch with the character and "live the part," rather than to the staged performance.

27. Moore, *The Stanislavski System,* p. 73.

28. This contention has subsequently been supported by numerous actors. See, for example, the interviews with Method actors in *Actors Talk About Acting,* L. Funke and J. Booth, eds. (New York: Random House, 1961).

29. Aristotle *Politics* 3.13.

30. Viśvanātha Cakravartin, *Rāgavartmacandrikā* 1.7. The edition I have used is one published in Bengali script, which was edited and translated into Bengali by Prāṇ Kiśor Gosvāmī (Howrah: Vinod Kiśor Gosvāmī, 1965).

31. Jīva Gosvāmin's commentary; see BRS 1.3.1.

32. The primary definition of the Rāgānugā Bhakti Sādhana is "imitation of the Vrajaloka" (*vrajalokānusāra*).

33. I refer the reader to Appendix A, a translation of the section of the *Bhaktirasāmṛtasindhu* in which Rūpa defines the Rāgānugā Bhakti Sādhana.

34. This idea is continuous with the Indian understanding of drama as expressed in Bharata's *Nāṭya-śāstra:*

> Bharata explains, in connection with the building of the theatre, how it is that the behaviour of the artist must of necessity be studied, and not impulsive; for the human actor, who seeks to depict the drama of heaven, is not himself a god, and only attains to perfect art through conscious discipline: "All the activities of the gods, whether in house or garden, spring from a natural disposition of the mind, but all the activities of men result from the conscious working of the will; therefore it is that the details of action to be done by men must be carefully prescribed."

See Ananda Coomaraswamy's introduction to *The Mirror of Gesture,* translated by Ananda Coomaraswamy and Gopala Duggirala (Cambridge: Harvard University Press, 1917), p. 3.

35. Jīva Gosvāmin, *Bhakti Sandarbha* 309, edited and translated into Bengali by Rādhāraman Gosvāmī Vedāntabhuṣan and Kṛṣṇagopal Gosvāmī (Calcutta: University of Calcutta, 1962), p. 538.

36. The *mādhurya-rasa* warrants a separate treatment by Rūpa in the *Ujjvalanīlamaṇi.*

37. For a good discussion of this, see Narendra Nath Law, "Śrī Kṛṣṇa and Śrī Caitanya," 2 parts, *The Indian Historical Quarterly* 23, no. 4 (December 1947): 261–99; and 24, no. 1 (March 1948): 19–66, pt. 2:278–84.

38. This is so because the Rāgātmikā *bhakta*s, the original Vrajaloka, are part of the essential nature of Kṛṣṇa (*svarūpa-śakti*). See the last section in Chapter 4.

39. These two options are discussed in greater detail in the following section of this chapter.

40. Jīva Gosvāmin, *Prīti Sandarbha* 82 ff. The edition I have used is one published in Bengali script, edited by Puridāsa Gosvāmin (Vṛndāvana: Haridāsa Śarmana, 1951).

41. This commentary is included in the Haridāsa Dāsa edition of the *Bhaktirasāmṛtasindhu.*

42. The ability to enter another's body is one of the traditional yogic powers. Patañjali's *Yoga-sūtras* 3.38 includes this power and calls it *para-śarīra-āveśa.* See also Maurice Bloomfield, "On the Art of Entering Another's Body: A Hindu Fiction Motif," *Proceedings of the American Philosophical Society* 56 (1917): 1–43.

43. Kuñjabihārī Dāsa, *Mañjarī-svarūpa-nirūpaṇa* (Rādhākuṇḍa: Ānanda Dāsa, 1975), p. 146. This Bengali work is a very informative source for the Rāgānugā Bhakti Sādhana as it is practiced in Vraja today.

44. *Caitanya-caritāmṛta* 2.8.222.

45. Rūpa Kavirāja, *Rāgānugāvivṛtti.* Although parts of this text were condemned for reasons which will be discussed in the following chapter, the presentation of identity transformation meets the approval of orthodox commentators then and now.

46. This text was published in Devanagari script with a Hindi translation by Haridāsa Śāstri (Vṛndāvana: Kṛṣṇadāsa Bābā, 1968). However, this publication

was an inexpensive local paperback which is difficult to attain. Since the text I used exists in manuscript form in the Vrindaban Research Institute (MS #1194), I will provide the Sanskrit when translating from this text.

abhimānas tridhā bṛhanmadhyamo laghur iti. bāhyadaśāyām satyām siddharūpe 'bhimāno laghuḥ, taṭastharūpe 'bhimāno bṛhat. arddhabāhyadaśāyām siddharūpe 'bhimāno madhyamaḥ, taṭastharūpe 'bhimāno madhyamaḥ. antardaśāyām siddharūpe 'bhimāno bṛhat, taṭastharūpe 'bhimāno laghuḥ. kevalāntar daśāyām siddharūpe 'bhimāno 'tibṛhat. kevalabāhyadaśāyām yathāsthitadehe 'bhimāno 'tibṛhat.

47. Dr. Acyuta Lal Bhatta, personal conversations in Vṛndāvana, 1981–82.

48. Detailed discussion of the actual practice of this visualization, *līlāsmaraṇa,* follows in Chapter 7.

49. Although the Sanskrit term *sevā* is usually translated as "service," I translate it as "performance," or "performative acts," to indicate its dramatic nature. *Sevā* is the action, modeled after the behavior of the exemplary Vrajaloka, by which the *bhakta* shows love for Kṛṣṇa.

50. Detailed discussion of the performative acts engaged in with these two bodies also follows in Chapter 7.

51. *tayā (sevayā) siddharūpe 'bhimānasyotkarṣaḥ sādhyate. . . . tayā (sevayā) yathāsthitadehe bādhitābhimānasya nāśaḥ sādhyate. pūrvābhimānaś cic-cchaktivṛttirūpaḥ paro 'bhimāno māyāvṛttirūpaḥ. pūrvasya sādhyatvāt, parasya nāśyatvāt.*

52. *Prīti Sandarbha. parikara-viśeṣabhimāninaḥ,* where *parikara* means the Vrajaloka.

53. See, for example, *Caitanya-bhāgavata* 1.9 (Gauḍīya Maṭh Edition).

54. See Moore, *The Stanislavski System,* pp. 27–28.

55. Kinsley, *The Divine Player,* p. 211.

56. Ibid., p. 145.

57. Ibid., p. 155.

58. De, *Vaiṣṇava Faith and Movement,* pp. 176–77. Emphasis mine.

59. Cakravarti, *Philosophical Foundations of Bengal Vaisnavism,* pp. 195–96 and 203. Emphasis mine.

60. O. B. L. Kapoor, *The Philosophy and Religion of Śrī Caitanya* (New Delhi: Munshriram Manoharlal Publishers, 1977), p. 196. Emphasis mine.

61. N. C. Ghose, Introduction to *Sree Gouranga Lilamritam* (Calcutta: Nitya Swarup Brahmachary, 1916), pp. xl–xli. Emphasis mine.

62. See, for example, the Bengali work by Rādhāgovinda Nāth, *Gauḍīya Vaiṣṇava Darśana,* 5 vols. (Calcutta: Pracyabani Mandir, 1957–60), 3 (1958): 2189 ff.

63. See Viśvanātha Cakravartin's commentary on *Bhaktirasāmṛtasindhu* 1.2.297. However, not all writers of this time follow his contention. Several commentators on the Rāgānugā Bhakti Sādhana use the term *anukāra* synonymously for *anuga* and *anusāra.* See, for example, the *Rāgānugāvivṛtti* of Rūpa

Kavirāja (early seventeenth century) and the *Sādhanadīpika* of Narāyana Bhaṭṭa (sixteenth century). Viśvanātha was perhaps reacting to developments in the Rāgānugā Bhakti Sādhana that involved what he considered a misuse of the notion of imitation (*anukāra*). See the following chapter for more on this issue.

64. See, for example, the Bengali translation and commentary on BRS 1.2.295 by Haridāsa Dāsa, and the Hindi translation and commentary of BRS 1.2.270 by Śyāmadāsa (Vṛndāvana: Harinām Press, 1981).

65. *Bhakti Sandarbha* 322, p. 547.

66. Donna M. Wulff, *Drama as a Mode of Religious Realization: The Vidagdhamādhava of Rūpa Gosvāmin* (Chico, Calif: Scholars Press, 1984), p. 32.

67. The idea was to "conform" (*conforme*) oneself to the Form, the paradigmatic life of Christ, through imitative activity.

68. The term "imitation" has been used in this positive sense by many social scientists and philosophers of religion (see n. 21, Chapter 1). The term *imitatio* also has had a very positive meaning among Christian monastic theologians, such as the Cistercians.

69. Wulff, *Drama as a Mode of Religious Realization,* p. 30.

70. Dimock, *Place of the Hidden Moon,* pp. 32 and 164. Emphasis mine.

71. Wulff, *Drama as a Mode of Religious Realization,* p. 29. Wulff bases her statement on claims such as this one by Joseph O'Connell: "It has been argued by Pran Kisor Gosvami that disapproval of *ahaṃgrahopāsana* ("thinking of oneself as the object of worship") also rules out identification of one's beatified body with that of a *gopī* whom Kṛṣṇa enjoys in sexual embrace since the *gopīs* too are considered in Gauḍīya Vaiṣṇava theology to be not different from Krishna himself" ("Social Implications of the Gauḍīya Vaiṣṇava Movement," pp. 242–43).

72. Shashibhusan Dasgupta, *Obscure Religious Cults* (Calcutta: Firma KLM, 1946), p. 125, cited by Wulff, *Drama as a Mode of Religious Realization,* pp. 29–30.

73. A. K. Majumdar, *Caitanya: His Life and Doctrine* (Bombay: Bharatiya Vidya Bhavan, 1969), p. 303.

74. Law, "Śrī Kṛṣṇa and Śrī Caitanya."

75. Wulff, *Drama as a Mode of Religious Realization,* p. 175.

76. A. K. Majumdar, for example, writes, somewhat ambiguously: "The developing thirst that follows in the wake of *kāmarūpā-bhakti* is called *kāmānugā,* which is of two kinds: in one called *sambhog-ecchāmayī,* the devotee is either a female or one who develops a feminine attitude of celestial amour with Kṛṣṇa; in the other; called *tad-bhāv-ecchātmikā,* the desire is for cultivating the emotions of love sentiment" (*Caitanya: His Life and Doctrine,* p. 303). In his discussion of Rāgānugā, S. K. De presents the two options in equally ambiguous terms. "Kāmānugā, which may again be either desire for enjoyment inspired by a sense of Kṛṣṇa's sport (*sambhogecchāmayī*), or a desire to realise those particular

Bhavas (*tattadbhāvecchātmikā*) of particular Gopīs, especially the erotic feeling (*bhāva-mādhurya-kāmitā*)" (*Vaiṣṇava Faith and Movement,* p. 179). More detail and clarity are needed to grasp the difference between these two options.

77. See Appendix A.

78. *Prīti Sandarbha* 365, p. 142.

79. Kuñjabihārī Dāsa, *Mañjarīsvarūpanirūpaṇa,* p. 25.

80. Viśvanātha Cakravartin, for example, echoes Jīva in his commentary on these verses. See his commentary, BRS 1.2.298–99.

81. Jīva Gosvāmin's commentary; see BRS 1.2.306.

82. Viśvanātha's commentary; see BRS 1.2.306. This is one of the essential differences between *bhakti* and Tantrism. In Tantrism the practitioner identifies with a deity, whereas in *bhakti* the practitioner tries to establish a relationship with a deity.

83. See above, p. 81. The *mañjarī* is discussed in the last section of the following chapter.

84. Jīva's commentary; see BRS 3.3.106.

85. Kuñjabihārī Dāsa, *Mañjarīsvarūpanirūpaṇa,* p. 37.

86. *Caitanya–caritāmṛta* 2.8.202–4 (Gauḍīya Maṭh edition).

87. This is why a friendship with Rādhā is more important for modern Gauḍīya Vaiṣṇavas than a friendship with Kṛṣṇa. In present-day Rādhākuṇḍa, one sees many of the *bābās* with the word "Rādhā" written on their bare chests with mud from the bottom of the Rādhākuṇḍa pond.

88. *Caitanya-caritāmṛta* 2.8.207–8.

89. See also A. K. Ramanujan, *Hymns for the Drowning* (Princeton: Princeton University Press, 1981), p. 162.

90. Entrance into Vṛndāvana is considered to be a great blessing. The inhabitants of Vṛndāvana never tired of telling me how fortunate I was to be in Vṛndāvana. They attributed this good fortune to meritorious action in a past life.

91. As one might expect, this lake is a favorite pilgrimage site for Rāgānugā practitioners in quest of a *gopī*-body. In fact, many serious *sādhus* coming to Vṛndāvana will bathe in this lake before entering the town.

92. Wendy D. O'Flaherty tells of a similar incident from the *Padma Purāṇa* involving Nārada. Nārada was only able to enter into Vṛndāvana and enjoy Kṛṣṇa after he had been transformed into a woman by bathing in a particular lake (*Dreams, Illusions, and Other Realities* [Chicago: University of Chicago Press, 1984], p. 83).

93. Some accounts, including that of the Bhaktimālā, have it that the Vṛndāvana Gosvāmin was Jīva, though he probably would have been too young at the time of Mīrā's visit. See A. J. Alston, *The Devotional Poems of Mirabai* (Delhi: Motilal Banarsidass, 1980), p. 6.

94. See also, *Caitanya-caritāmṛta* 2.8.230–31.

95. Kṛṣṇa's kindness, however, is immense. The story continues to tell how Kṛṣṇa went to Lakṣmī out of compassion. Today one sees an image of Kṛṣṇa beside that of Lakṣmī in the temple at Bel-ban.

96. Kṛṣṇadāsa Kavirāja says in the *Caitanya-caritāmṛta* that one cannot attain Kṛṣṇa in Vraja without imitating the *gopīs* in a *siddha-deha* (2.8.229–30).

97. *Siddha-rūpa* (or *siddha-deha*) can be translated as the "perfected body," but is best left untranslated as a technical term.

98. Jīva's commentary; see BRS 1.2.295.

99. See Dasgupta, *Obscure Religious Cults,* pp. 219, 228, and 254–55.

100. Dasgupta explains that for the *yogīs* this body can disappear into a higher state of liberation and attain a form of pure light (ibid., p. 255). This notion of the *siddha-deha* is much closer to the Gauḍīya Vaiṣṇava understanding of the term.

101. See Chapter 7 for a discussion of the *līlā-smarana* meditation.

102. *Prīti Sandarbha* 10, p. 15.

103. Ibid., 10–11, p. 15. *Bhāgavata Purāṇa* 3.15.14 and 1.6.29.

104. *Rāgavartmacandrikā* 2.7.

105. *Padma Purāṇa* 83.7–8. Niradprasād Nāth, however, points out the fact that, although this section of the *Padma Purāṇa* so well fits the Gauḍīya Vaiṣṇava theory of Rāgānugā, it was not quoted by the early Gosvāmins who knew the text well. This suggests that this section may have been added after the time of the Vṛndāvana Gosvāmins. See Niradprasād Nāth, *Narottama Dāsa o Tāhār Racanāvali* (Calcutta: University of Calcutta, 1975), p. 335. I am grateful to Tony K. Stewart for bringing this study to my attention.

106. A late sixteenth- early seventeenth-century text describing the *līlā* of Rādhā and Kṛṣṇa and the meditation on this *līlā.* The *Gaura-Govindārcana-smaraṇa-paddhati* has been published in Bengali script in a collection of three such texts: *Paddhati-trayam,* Haridāsa Dāsa, ed. (Navadvīpa: Haribol Kutir, 1948).

107. Ibid., verses 303–312, pp. 41–42.

108. Most Rāgānugā practitioners still follow this elevenfold schema today. The eleven aspects of the *siddha-rūpa* were repeated by Dhyānacandra (in his *paddhati*s, published in the above volume), except that beauty (*rūpa*) was changed to read complexion (*varṇa*). I interviewed some Rāgānugā practitioners in the Vraja region who used a twelvefold schema to define the *sidda-rūpa,* involving: name, age, ornaments, complexion, color of sari, disposition, speciality, service, residence, *yoga-pītha,* bower, and main *mañjarī.*

109. I take this list from N. N. Law, "Śrī Kṛṣṇa and Śrī Caitanya," 2:47. Some of the details he took from Kavikarṇapūra's *Gaura-ganoddeśa-dīpikā.* This list is in agreement with most of the *yoga-pītha*s I observed in Vraja. The *Mānasī Sevā* repeats many of these aspects and adds the name of Rūpa Mañjarī's mother, Sumanā, and mother-in-law, Jaṭilā. The manuscript of this text resides in the Vrindaban Research Institute. See Tony K. Stewart, "The Biographical Images of Kṛṣṇa–Caitanya: A Study in the Perception of Divinity" (Ph.D. dissertation, University of Chicago, 1985), Appendix F, pp. 499–500, for a list of the features of the *siddha-rūpas* of all six of the Vṛndāvana Gosvāmins.

110. Some Bengali practitioners have dual *siddha-rūpa*s. One is a *gopī* in

Vraja serving Rādhā and Kṛṣṇa, and the other is a *brahmin* boy in Navadvīpa serving Caitanya. See Rādhāgovinda Nāth, *Gauḍīya Vaiṣṇava Darśana* 3: 2214. I mention this in passing, but will restrict my treatment to the *gopī* body of Vraja, since this is certainly the most common view of the *siddha-rūpa* found in the Gauḍīya Vaiṣṇava texts and held by practitioners in Vraja today.

111. Rādhāgovinda Nāth mentions the different types of potential *siddha-dehas* in current practice (*Gauḍīya Vaiṣṇava Darśana* 3: 2193).

112. A spring celebration where the participants throw colored powder and spray colored water on each other. For an interpretive description of Holi, see McKim Marriott, "The Feast of Love," in Milton Singer, ed., *Krishna: Myths, Rites, and Attitudes* (Chicago: University of Chicago Press, 1968), pp. 200–212.

113. This seems to be a reversal of the *advaitin* texts described by Wendy O'Flaherty, in which *līlā* time is usually much longer than ordinary time. See *Dreams, Illusions and Other Realities,* especially pp. 132–134.

114. Rasika Mohana Vidyabhusana, *Śrīmat Dāsa Gosvāmin,* pp. 156–57, cited by N. N. Law, "Śrī Kṛṣṇa and Śrī Caitanya," 2: 51–52.

115. See Stephan Beyer, *The Cult of Tara* (Berkeley: University of California Press, 1973), pp. 82–92.

116. Rūpa Kavirāja, *Rāgānugā-vivṛtti. bhāvanāmayam taptam . . . satyam dagdhāṅguṣṭhatayā bahiḥ pīḍāyāḥ satyatvāt.*

Chapter 6

1. A translation of the first verse of poem 45 of the Prārthanā Padāvali of Narottama Dāsa Thākura in Nāth, *Narottama Dāsa o Tāhār Racanāvali,* p. 343–44.

2. See Paul Levy, *Buddhism: A Mystery Religion?* (London: University of London, Athlone Press, 1957), pp. 1–37. An odd thesis, but a good presentation of Cambodian monastic ordination.

3. See Yoshito S. Hakeda, *Kūkai: Major Works* (New York: Columbia University Press, 1972), p. 98.

4. See Bernard of Clairvaux, *The Steps of Humility and Pride,* translated by M. Ambrose Conway (Kalamazoo: Cistercian Publications, 1980); and Ettienne Gilson, *The Mystical Theology of Saint Bernard,* translated by A. H. C. Downes (New York: Sheed & Ward, 1940).

5. See Marta Weigle, *Brothers of Light, Brothers of Blood: The Penitentes of the Southwest* (Albuquerque: University of New Mexico Press, 1976).

6. See John G. Neihardt, *Black Elk Speaks* (New York: Pocket Books, 1959), pp. 136–47.

7. Rādhāgovinda Nāth, *Gauḍīya Vaiṣṇava Darśana,* 3:2193.

8. Jīva Gosvāmin, *Bhakti Sandarbha* 309 p. 540. Based on *Bhāgavata Purāna* 7.5.23.

9. Kṛṣṇadāsa Kavirāja, *Caitanya-caritāmṛta* 2.22.156–57.

10. Plenty of evidence exists to demonstrate that some practitioners did and

still do follow this more literal path. This is discussed in greater detail in the section of next chapter entitled "Physical Role-Taking."

11. There may be one exception. Dimock mentions a man named Rūpa Kavirāja who was involved in negative reactions to an incident where the Govardhana stone belonging to Raghunātha Dāsa was given to two women (see *Place of the Hidden Moon,* p. 100). Whether this is the same Rūpa Kavirāja or not is uncertain.

12. More information on Rūpa Kavirāja may exist in the manuscript collection of a small temple in Vṛndāvana which is maintained by some of his followers. I was unable to examine these documents. Additional information might also exist in the remote *āśrama* in Assam where he spent his last days.

13. Haridāsa Dāsa, *Gauḍīya Vaiṣṇava Abhidhāna,* 3 vols. (Navadvīpa: Haribol Kuṭir, 1957), 3: 1350. (A Bengali dictionary of Gauḍīya Vaiṣṇavism.)

14. Krishnagopal Goswami, "Introduction to Rūpa Kavirāja," *Sārasaṅgraha,* Krishnagopal Goswami, ed. (Calcutta: University of Calcutta, 1949), pp. xlii–iii.

15. See Haridāsa Dāsa's footnote to Viśvanātha's commentary on verse 1.2.295 in his Bengali translation of *Bhaktirasāmṛtasindhu,* pp. 138–39.

16. Rūpa Kavirāja, *Sārasaṅgraha,* Krishnagopal Goswami, ed. (Calcutta: University of Calcutta, 1949). When the *Sārasaṅgraha* was published, the editor had much doubt as to its author. Even though several scholars had attributed it to Jīva Gosvāmin, the editor leaned toward Rūpa Kavirāja in his final decision, but admitted that "it is difficult to identify the author." His uncertainty caused him to print the text as "attributed to Rūpa Kavirāja." I think, however, it is almost certain that the *Sārasaṅgraha* was written by Rūpa Kavirāja, and not by Jīva Gosvāmin. My evidence is twofold. First the terminology, style, and theory expressed in the *Sārasaṅgraha* are identical to that of the *Rāgānugāvivṛtti,* a text certainly written by Rūpa Kavirāja. Second, both the *Sārasaṅgraha* and the *Rāgānugāvivṛtti* are recorded as being written by Rūpa Kavirāja in the document that emerged from the Jaipur council condemning these works. This Hindi document was reproduced by Nareścandra Bansal, *Caitanya Sampradāya: Siddhānta aur Sāhitya* (Agra: Vinod Pustaka Mandir, 1980), pp. 504–6.

Rāgānugāvivṛtti, Vrindaban Research Institute, MS# 1192–95 (I cite MS# 1194). Another manuscript was printed by a local Vṛndāvana press: *Rāgānugāvivṛtti,* edited and translated into Hindi by Haridāsa Śāstri (Vṛndāvana: Kṛṣṇadāsa Bābā, 1968). This edition is full of errors, but is useful if one can locate it. I read the *Rāgānugāvivṛtti* in Vṛndāvana with Dr. Acyuta Lal Bhatta and again in Chicago with Neal Delmonico. I an grateful to both for their time and helpful comments.

17. Rūpa Kavirāja, *Sārasaṅgraha,* pp. 149–61.

18. Ibid., p. 151.

19. Ibid., p. 78.

20. *Taṭastha-rūpa* can be translated as the "neutral body." It refers to the body that has not yet engaged in *sādhana.* In the context of Gauḍīya Vaiṣṇavism it may also refer to the body attached to the *taṭasthā-śakti,* the source of the

human position lying halfway between the essence of God (*svarūpa-śakti*) and the world of illusion (*māyā-śakti*).

21. Rūpa Kavirāja, *Sārasaṅgraha,* p. 78. In the *Rāgānugāvivṛtti* Rūpa comments that "the words 'attached to thinking of oneself as a male,' " used to describe the *taṭastha-rūpa,* "were used because maleness is the dominant sex, but this is really to be understood to mean the 'body as it is' (*yathāsthita-deha*)." *sva-puṃstva-bhāvanā-yugityatra puṃstva-pradhānyād-eva, kintu yathā-sthita-deha-mātratvam jñeyam.*

22. *Rāgānugāvivṛtti.* The first: *antar-bhūta-yathā-sthita-rūpāprāpta-śrī-kṛṣṇādi-saṅga-vraja-janānukāri-bhāvanā-maya-rūpa;* and the second: *antar-bhūta-prāpta-śrī-kṛṣṇādi-saṅga-vraja-janānukāri-bhāvanā-maya-rūpa-yathā-sthita-rūpa.*

23. This initiation is described in detail in the following chapter.

24. *Rāgānugāvivṛtti. sādhaka-rūpe vraja-janānukāri-bhāvanā-maya-rūpasyāropaḥ.* An additional anonymous Bengali text, entitled *Rāgānugā Sādhana,* which is part of the collection of the Vrindavana Research Institute (MS# B 315), and seems to be based on the *Rāgānugāvivṛtti,* puts it this way: "We will put the meditative body inside the ordinary body." (*yathā-sthita-deher bhāvanā-maya-dehir antar-gata karibo.*)

25. *Rāgānugāvivṛtti. taṭa-stha-rūpasya sādhaka-rūpaṃ nāvasthā, kintu dehāntaram.*

26. Ibid., *yathā kāñcanatām yāti kāmsyam rasa-vidhānataḥ, tathā dīkṣā-vidhānena dvijatvaṃ jāyate nṛṇam iti vacanād vaiṣṇava-mantra-grahaṇena dehāntaram syāt.*

27. Ibid., *sādhaka-rūpe rāgānugātaḥ sādhaka-rūpe 'ntar-daśā sādhaka-rūpe vraja-janānukāri-bhāvanā-maya-rūpasyāropāt.*

28. *Sārasaṅgraha,* p. 151.

29. *Rāgānugāvivṛtti. ataḥ sādhaka-rūpasya cātur-vidham. cātur-vidhye krameṇa bāhya-daśā'ntar-bāhya-daśā pūrvāntar-daśā parāntar-daśā.*

30. Ibid., *vīṇā-vaiṇika-tulyatvam siddha-sādhaka-rūpayoḥ. yathā tayor vibhede 'pi gānayor eka-tānatā tulya-sva-rūpavattvañ ca; tathā tad-rūpayor dvayoḥ bhede 'pi sevayoḥ sāmyam eka-kālatvam eva ca. yathā vaiṇika-citta-sthā gītir vīṇā-bhavā; tathā sevā sādhaka-rūpa-sthā siddha-rūpe pravarttate. bhede yathā raso na syād vīṇā-vaiṇika-gānayoḥ; tathā tat-sevayor bhede vraja-bhāvo na jāyate.*

31. Ibid., *ata eva siddha-rūpeṇa yat karoti sādhaka-rūpe tat pracarati; sādhaka-rūpeṇa yat karoti siddha-rūpe tat pracarati. tasmāt siddha-sādhaka-rūpabhyām vraja-lokānusāra eva mano-vāk-kāyaiḥ kartavya iti.*

32. *Sārasaṅgraha,* p. 157.

33. Ibid., *yathā sādhaka-rūpeṇa kriyamāṇā sevā anyair janair yathā-sthita-dehena kriyamāṇā pratīyate, tathā varṇāśrama-rahitena dehena kriyamāṇā sevā anyair janair varṇāśramavatā dehena kriyamāṇā pratīyate.*

34. Ibid., *sādhaka-rūpa-siddha-rūpa-gatāntar-daśāyor vailakṣaṇyam yathā-sthita-rūpa-yogāyogabhyām.*

35. See Bansal, *Caitanya Sampradāya,* pp. 504–506.

36. One hopes this does not apply to curious historians.

37. For the life and works of Viśvanātha, see Haridāsa Dāsa, *Gauḍīya Vaiṣṇava Abhidhāna,* 3: 1370.

38. Local tradition in Vṛndāvana has it that Viśvanātha was a reincarnation of Rūpa Gosvāmin.

39. Viśvanātha's commentary is included in the Haridāsa Dāsa edition of the *Bhaktirasāmṛtasindhu.*

40. Viśvanātha Cakravartin, *Rāgavartmacandrikā;* the edition I cite is one published in Bengali script, which was edited and translated into Bengali by Prāṇ Kiśor Gosvāmin (Howrah: Vinod Kiśor Gosvāmin, 1965). This text has been translated into English by Joseph O'Connell, "Rāgavartmacandrikā of Viśvanātha Cakravartin," in *A Corpus of Indian Studies,* A. L. Basham et al., eds. (Calcutta: Sanskrit Pustak Bhandar, 1980). The translation is good, but O'Connell disappointingly translates Rāgānugā Bhakti as "passionate devotion." The notion of imitation (*anuga*) is absent from such a translation.

41. *Rāgavartmacandrikā* 2.8., p. 38.

42. Ibid., 1.12., p. 11.

43. Ibid., 1.11., p. 8.

44. The word *mañjarī* itself is somewhat of a mystery. Monier-Williams defines it as a "flower, bud, or shoot." One of the most convincing definitions I heard while in Vṛndāvana is that it is a stamen, that part of a flower closest to its center. Considering the frequent conception of the stage of the Vṛndāvana-līlā as a lotus flower, this makes good sense. Rādhā and Kṛṣṇa together are the center, the *sakhī*s are the petals, and the *mañjarī*s are the stamens.

45. I do not mean to give the impression that the other roles delineated by Rūpa did not survive. One still finds practitioners in Vraja today who follow the role of a male companion (*sākhya-bhāva*), the role of an elder (*vātsalya-bhāva*), as well as the other optional roles within the *mādhurya-bhāva.*

46. See second part of third section of Chapter 4.

47. Kuñjabihārī Dāsa writes: "Among the five types of *sakhī*s, the *prāṇa-sakhī* and the *nitya-sakhī,* who have a great love for Rādhā, are called by the name *mañjarī*" (*Mañjarīsvarūpanirūpaṇa,* p. 39).

48. Our culture agrees that the adolescent female is emotionally the most intense, but being a culture that basically distrusts the emotions, ours has not held the adolescent female as a religious ideal.

49. These examples are found in the prayers of Narottama Dāsa. See Nirodprasād Nāth, *Narottama Dāsa o Tāhār Racanāvali,* pp. 307–53.

50. *Caitanya-caritāmṛta* 2.8.212–13. Here Rādhā assumes the role of a *sakhī.*

51. See the introduction to Nāth, *Narottama Dāsa o Tāhār Racanāvali,* pp. 93–123.

52. For a brief sketch of the life and works of Kavikarṇapūra, see De, *Vaiṣṇava Faith and Movement,* pp. 41–45. The edition of the *Gaura-gaṇoddeśa-dīpikā* I cite is one included in a collection of texts edited with a Hindi translation

by Kṛṣṇadāsa Bābā, *Grantha-ratna-pañcakam* (Mathurā: Kṛṣṇadāsa, Pusparāja Press, 1953).

53. *Gaura-gaṇoddeśa-dīpikā* 180–87, pp. 34–35.

54. A *yogapīṭhāmbuja* is a lotus-like diagram inscribed with the *siddha* names of the main figures of the līlā. Dhyanacandra's *yogapīṭha* has been reproduced by Haridāsa Dāsa, *Gauḍīya Vaiṣṇava Abhidhāna* 1: 633. See also Figure 3.

55. The *Rāgamālā* is included in Niradprasād Nāth, *Narottama Dāsa o Tāhār Racanāvali,* pp. 633–43.

56. See *Bhaktirasāmṛtasindhu* 3.5.7–8.

57. Kuñjabihārī Dāsa, *Mañjarī-svarūpa-nirūpaṇa,* p. 57.

58. This shift from Kṛṣṇa as the *viṣaya* to Rādhā-Kṛṣṇa as the *viṣaya* interestingly parallels the shift from the early notion that Caitanya was an incarnation of Kṛṣṇa to the notion expressed in Kṛṣṇadāsa Kavirāja's *Caitanya-caritāmṛta* that Caitanya was a dual incarnation of Rādhā-Kṛṣṇa. For more on this latter shift, see Stewart, "The Biographical Images of Krsna-Caitanya: A Study in the Perspective of Divinity," chapters 5–6.

59. *Bhāvollāsa* literally means "joyful emotion," but is the technical term for a particular kind of *sakhī*'s emotion and is best left untranslated.

60. Gopāla Gosvāmin, *Gaura-Govindārcana-smaraṇa-paddhati* 305, p. 41.

61. Kuñjabihārī Dāsa, *Mañjarī–svarūpa–nirūpaṇa,* pp. 61–62. I read this section of this text in Chicago with Prof. Edward Dimock and Neal Delmonico.

62. Gopālaguru Gosvāmin, for example, defines the *mañjarī* as one who imitates Rūpa Mañjarī (*rūpa-mañjary-anugatā*) (*Gaura-Govindārcana-smaraṇa-paddhati* 307, p. 41). Also Narottama makes frequent reference to the fact that the *mañjarī* is to follow Rūpa Mañjarī. See the *Sādhana-candrikā* and the *Sādhana-bhakti-candrikā,* texts included in Nāth, *Narottama Dāsa o Tāhār Racanāvali,* pp. 447–63 and pp. 705–12.

63. The ideal *guru* of Gauḍīya Vaiṣṇavism should be one such exemplary *siddha.*

CHAPTER 7

1. A *kuṇḍa* is a pond, thus Rādhā's pond.

2. The tradition has it that Caitanya himself discovered this pond on his pilgrimage to Vraja. According to Fredrick Growse, the sandstone steps were added in the early nineteenth century by Krishnan Chandra Sinh, better known as Lālā Bābu (*Mathurā: A District Memoir* [New Delhi: Asian Educational Services, 1882], p. 258).

3. Rūpa Gosvāmin, *Upadeśāmṛta* 9–11.

4. The power of the *mantra* is expressed, for example, in the second half of the *Bṛhad–bhāgavatāmṛta* of Sanātana Gosvāmin.

5. See Jīva Gosvāmin, *Bhakti Sandarbha* 202–7, pp. 345–52. See also Rādhāgovinda Nāth, *Gauḍīya Vaiṣṇava Darśana,* 3:2238–78; and Sundarānandadāsa Vidyāvinod, *Vaiṣṇava Siddhānte Śrīgurusvarūpa* (Calcutta: Karuṇā Dāsa, 1964).

6. Kṛṣṇadāsa Kavirāja, for example, claimed to have had six *śikṣā-gurus*: the six Vṛndāvana Gosvāmins (*Caitanya-caritāmṛta* 1.1.36).

7. See Jīva Gosvāmin, *Bhakti Sandarbha* 208, p. 353.

8. This is a text written by Gopāla Bhaṭṭa which contains procedures for various rituals. For a brief description of the initiation outlined by this text, see De, *Vaiṣṇava Faith and Movement,* pp. 455–60. One wonders, however, if the actual initiation was ever as elaborate as this literary ideal.

9. *Hare Kṛṣṇa Hare Kṛṣṇa, Kṛṣṇa Kṛṣṇa Hare Hare, Hare Rāma Hare Rāma, Rāma Rāma Hare Hare.*

10. This new body is the *siddha-rūpa,* not the *sādhaka-rūpa* of Rūpa Kavirāja.

11. Jīva Gosvāmin, *Bhakti Sandarbha* 283, p. 484.

12. Rādhāgovinda Nāth, *Gauḍīya Vaiṣṇava Darśana* 3: 2285.

13. See the fourth section of Chapter 5.

14. Gopālaguru Gosvāmin, *Gaura–govindārcana-smaraṇa-paddhati* 309, p. 42.

15. A small pond located in Vraja.

16. O. B. L. Kapoor, *Śrī Rāmadāsa Bābājī* (Vṛndāvana: Śrī Rādhā Govinda Press, 1982), pp. 114–15.

17. Kuñjabihārī Dāsa, *Mañjarī svarūpa nirūpaṇa,* pp. 145–46.

18. A *yoga-pīṭhāmbuja* is literally a "lotus which is the place of union." Perhaps it is a remnant of the eight-petaled lotus which was part of the *dīkṣā-maṇḍala* mentioned in the elaborate initiation of the *Haribhakti-vilāsa.* See De, *Vaiṣṇava Faith and Movement,* p. 456.

19. This particular *yoga-pīṭhāmbuja* is a reproduction of several *yoga-pīṭhāmbujas* I copied while in Vṛndāvana and Rādhākuṇḍa.

20. A discussion of these two options is found in Gauragovindānanda Bhāgavatasvāmī's *Rāgānugā-bhakti-tattva-kusumāñjali* (Calcutta: Classic Press, 1948), pp. 22–23.

21. See the first section of Chapter 5.

22. Wendy O'Flaherty concludes her fascinating study of the mental worlds of India with this observation: "In India, the realm of mental images is not on the defensive. . . . In India the dream that wanders in the daylight does not fade but instead makes daylight all the more luminous; it shines into the hidden corners of waking life to show us shadows brighter than the light" (*Dreams, Illusion and Other Realities,* p. 304).

23. Rāmānuja, *Brahma Sūtras, Śrī Bhāṣya* 1.1.

24. Ibid., 1.1.

25. *evam rūpa dhrūvānusmṛtir eva bhakti-śabdenābhidhīyate.* Ibid., 1.1.

26. Rāmānuja, *Gītā-bhāṣya* 18.65.

27. See Robert C. Lester, *Rāmānuja on the Yoga* (Madras: Adyar Library and Research Centre, 1976), pp. 140–41.

28. Jīva Gosvāmin, *Bhakti Sandarbha* 278, pp. 475–76.

29. Viśvanātha Cakravartin, for example, maintains that there is a preeminence of *smaraṇa* in Rāgānugā. *tatra rāgānugāyām smaraṇasya mūkhyatvam.* Viśvanātha Cakravartin, *Bhaktirasāmṛtasindhu-bindu.* The edition I cite is one

published in Devanagari with a Hindi translation and commentary by Śyāma Dāsa (Vṛndāvana: Harinām Press, 1977), p. 128.

30. The *aṣṭa-kālīya-līlā smaraṇa* meditation has received little attention by scholars. One exception is a paper by Neal Delmonico, "Time Enough for Play: Religious Use of Time in Bengal Vaiṣṇavism," paper presented at the Bengal Studies Conference, Chicago, June 1982.

31. These three traditional dramatic terms are used, for example, in a short text on Mañjarī Sādhana written by Narottama Dāsa. See the *Sādhana–candrikā* in Nath, *Narottama Dāsa o Tāhār Racanāvali,* p. 468. Narottama's texts also provide elaborate descriptions of such meditative scenes. See, for example, the *Kuñjavarṅana,* which provides a vivid picture of the site of the Rādhākuṇḍa *līlā.*

32. Gopālaguru Gosvāmin, *Śrī-Rādhā-Kṛṣṇayor Aṣṭa-kālīyā-līlā-smaraṇa-krama-paddhati,* included in *Paddhati-trayam,* Haridāsa Dāsa, ed., p. 70. A *muhūrta* is a period of forty-eight minutes. I am grateful to Neal Delmonico for pointing these verses out to me. An edition of the *Smaraṇa-maṅgala-stotram* was published by Kṛṣṇadāsa Bābā (Govarddhana, n.d.)

33. See Appendix B for a translation of this poem. The Sanskrit appears in De, *Vaiṣṇava Faith and Movement,* pp. 673–75.

34. Niradprasād Nāth, *Narottama Dāsa o Tāhār Racanāvali,* pp. 114–15.

35. Edwin Gerow points out that the Vaiṣṇavas were the first to sign their poetry. "Rasa as a Category of Literary Criticism," p. 242.

36. The text I used has been published with a Hindi translation by Haridāsa Śāstri (Vṛndāvana: Śrīgadādhara Gaurahari Press, 1977–81, 3 vols.).

37. The most popular *guṭikā* used in Vraja today for the *līlā-smaraṇa* meditation is the *Gaura-govinda Līlāmṛta Guṭikā* of Siddha Kṛṣṇadāsa Bābā of Govardhana (Rādhākuṇḍa: Gopāladāsa, 1951).

38. William James, *Varieties of Religious Experience,* p. 335.

39. Niradprasād Nāth, *Narottama Dāsa o Tāhár Racanāvali,* pp. 342–43.

40. The practitioner of *bhakti,* however, never identifies with Kṛṣṇa. This fact distinguishes *bhakti* from Tantric visualization and identification, where the practitioner does identify with the god.

41. Kṛṣṇadāsa Kavirāja, *Caitanya-caritāmṛta* 2.22.156.

42. The early Bhāgavatas outlined five emanations of the Supreme Lord. For the purpose of worship and meditation, the Supreme appears in these different manifestations: supreme form (*parā*), powers (*vyūha*), indweller in the heart of all (*antaryāmin*), incarnation (*vibhava*), and the image (*arca*). The image is considered to be the most accessible to the worshipper. For more information about Hindu image worship, see Joanne Punzo Waghorne, and Norman Cutler, eds., *Gods of Flesh, Gods of Stone* (Chambersburg, Pa.: Anima Publications, 1985), particularly Vasudha Narayana, "Arcāvatāra: On Earth as He Is in Heaven," pp. 53–67.

43. The ritual is frequently called the *prāṇa-pratiṣṭhā,* in which the life-breath (*prāṇa*) is infused into the image. This ritual also often involves opening the eyes of the image. See Diana Eck, *Darśan: Seeing the Divine Image in India* (Chambersburg, Pa.: Anima Publications, 1981), pp. 39–41.

44. This is the mountain Kṛṣṇa lifted up and used as an umbrella to protect his village from the raging storm of Indra. The whole mountain is considered to be a manifestation of Kṛṣṇa, and thus stones from it do not need to be "installed."

45. This group must be distinguished from other groups of men in India who dress as women for any variety of reasons, many not religious.

46. Some are even trying to transform the physical body, à la Rūpa Kavirāja.

47. H. H. Wilson, *The Religious Sects of the Hindus* (Calcutta: Susil Gupta, [1861] 1958), p. 101. It is difficult to know exactly what the relationship between these Sakhī Bhāvas and the Gauḍīya Vaiṣṇavas might be. I seriously doubt if Wilson knew. He claims that they are associated with the Rādhā Vallabhīs, a sect closely related to the Gauḍīya Vaiṣṇavas.

48. Kinsley, *The Divine Player,* p. 172.

49. R. G. Bhandarkar, *Vaiṣṇavism, Śaivism and Minor Religious Systems* (Varanasi: Indological Book House, 1965), p. 86.

50. See the Bengali biography by Dineśacandra B. Gītāratna, *Śrī Lalitā Sakhī* (Calcutta: Nimaicandra Bhaṭṭācārya, 1942).

51. The *sahajiyā* option. See Dimock, *Place of the Hidden Moon,* especially the chapters on *sādhana.* Also there is some evidence that some *guru*s of the Vallabhācārya sect engaged in sexual acts with their female devotees, although after examining the evidence that was presented in the British courts it seems to me that much of the affair was a classic example of cross-cultural misunderstanding and colonial misrepresentation. Nevertheless, the claims were made: "I was aware that the females of my sect believed the Maharajas to be incarnations of Krishna, and that as the gopis obtained salvation by falling in love with Krishna, our females were bent upon adulterous love towards the Maharajas" (Karsandas Mulji, *History of the Sect of Maharajas, or Vallābhācaryas, in Western India* [London: Trubner & Co., 1865], Appendix, "Specimens of the Evidence and the Judgement in the Libel Case," p. 17).

52. Pilgrims entering Vṛndāvana will often exclaim that now they are truly complete, for they have set foot in the auspicious land of Vraja.

53. Rūpa Gosvāmin, *Upadeśmṛta* 9. The edition I use is one published in Bengali script, which was edited with a Bengali translation by Bhaktikevala Auḍulomi Mahārāja (Calcutta: Śrī Gauḍīya Sampradāya, 1980).

54. The region of Vraja is variously called Mathurā-maṇḍala, Vraja-bhūmi, Vraja-dhāma, or simply Vraja.

55. Government statistics today show that more people visit Vṛndāvana each year than the Taj Mahal at Agra, though the reasons for the respective visits are quite different and involve different types of people.

56. The best works on the *vana-yātrā* are in Hindi. See Seṭha Govindadāsa, *Vraja aur Vraja-yātrā* (Delhi: Bhāratīya Viśva Prakāśana, 1959); and P. D. Mital, *Vraja kā Sanskritika Itihāsa* (Delhi: Rajakamal Prakashan, 1966).

57. Charlotte Vaudeville, "Braj, Lost and Found," *Indo-Iranian Journal* 18 (1976): 196.

58. *Bhāgavata Purāṇa* 10.47.59. See Cunnīlāl Śeṣa, "Vraja-yātrā kī Paraṃparā," in *Vraja aur Vraja Yātrā,* pp. 91–92.

59. See F. S. Growse, *Mathurā: A District Memoir,* p. 75. Norvin Hein agrees with Growse, relying on evidence from Hindi scholarship on the biographies of Nārāyana Bhaṭṭa (*The Miracle Plays of Mathura,* pp. 226–27). The Gauḍīya Vaiṣṇava scholar Haridāsa Dāsa also puts his weight behind this position (*Gauḍīya Vaiṣṇava Abhidhāna* 3: 1272).

60. Sesa, *Vraja aur Vraja Yātrā,* p. 108.

61. Perhaps Nārāyana Bhaṭṭa was not identified as a Vṛndāvana Gosvāmin because Caitanya did not send him to Vraja. Also the notion that he was an incarnation of Nārada places him in a different mythological circle than the Gosvāmins, who were seen as incarnations of *gopī*s. S. K. De does not discuss him in his major study, but only mentions in passing that Nārāyana Bhaṭṭa was cited in the *Sat-kriyā-sāra-dīpikā* for his work on Smarata rules (*Vaiṣṇava Faith and Movement,* pp. 138 and 531).

62. Jānakīprasāda, *Nārāyana Bhaṭṭa-caritāmṛta.* This text has been edited with a Hindi translation by Kṛṣṇadāsa Bābā (Mathurā: Kṛṣṇadāsa Bābā, 1955).

63. For more on *rāsa-līlā*s, see Hawley, *At Play with Krishna,* and Hein, *The Miracle Plays of Mathurā.*

64. Northrop Frye, *Anatomy of Criticism* (Princeton: Princeton University Press, 1971), p. 282.

65. Geertz's sense of the term "model" is intended. See Geertz, "Religion as a Cultural System," p. 93.

66. Hein, *The Miracle Plays of Mathurā,* p. 14.

67. Ibid., pp. 159–60.

68. Geertz, "Religion as a Cultural System," p. 112.

CHAPTER 8

1. This is, for example, the position of the Southern (Teṅgalai) or Cat-hold (Mārjāra-nyāya) School of Śrī Vaiṣṇavism. See Robert C. Lester, "Rāmānuja and Śrī-Vaiṣṇavism: The Concept of Prapatti or Saraṇāgati," *History of Religions* 5, no. 2 (1966): 266–82. This particular school of Śrī Vaiṣṇavism is the probable source of Rudolf Otto and Nathan Söderblom's ideas of *bhakti.*

2. See Robert C. Lester, "Aspects of the Vaiṣṇava Experience: Rāmānuja and Pillai Lokācārya on Human Effort and Divine Grace," *Indian Philosophical Annual* 10 (1974–75): 1–10.

3. *Bhaktirasāmṛtasindhu* I.3.6.

4. Rūpa's position is much closer to Rāmānuja's than to the later Śrī Vaiṣṇava writer Pillai Lokācārya, who promotes the *prapatti* path of unmerited grace.

5. I have also indicated throughout this study that one of the major differences between *bhakti* and Tantric identification is that, while the Tantric practitioner regularly identifies with the god, the *bhakti* practitioner identifies with a companion of the god. See also Wendy D. O'Flaherty, *Women, Androgynes and Other Mythical Beasts* (Chicago: University of Chicago Press, 1980), pp. 87–88.

6. Staged dramas depicting the Kṛṣṇa stories were used as a powerful me-

dium of propagation by the Vaiṣṇavas. The plays of the Assamese Vaiṣṇava leader Śaṅkaradeva (1449–1568) provide perhaps the clearest example of this. He wrote and directed many plays for the propagation of the Kṛṣṇa stories in an area where they were previously unknown. See Maheswar Neog, *Śaṅkaradeva and His Times: Early History of the Vaisnava Faith and Movement in Assam* (Gauhati: Gauhati University Press, 1965). This somewhat common genre of drama is identified by Northrop Frye as the "scriptural play" or "myth play," since the scripture or myth defining the ideal world of a tradition forms the script of the play (*Anatomy of Criticism*) [Princeton: Princeton University Press, 1971], p. 282).

7. "Holy actor" is Jerzy Grotowski's term. He says that the "holy actor" is one who uses a "role as a trampoline, an instrument with which to study what is hidden behind our everyday mask—the innermost core of our personality" (*Towards a Poor Theatre,* pp. 34–39). I borrow the term and add to it my own meaning: The "holy actor" is a monk-like religious imitator who strives to enact a transcendent role defined by a paradigmatic individual.

8. For a good presentation of Cistercian theology, see Ettienne Gilson, *The Mystical Theology of Saint Bernard,* translated by A. H. C. Downes (New York: Sheed & Ward, 1940).

9. See John G. Neihardt, *Black Elk Speaks* (New York: Pocket Books, [1932] 1972).

10. See *The Liturgical Sermons of Guerric of Igny,* translated by the monks of Mount Saint Bernard Abbey (Spencer, Mich.: Cistercian Publications, 1970).

11. See Paul Levy, *Buddhism: A Mystery Religion?* (New York: Schocken Books, 1957), pp. 1–37.

12. See John C. Holt, *Discipline: The Canonical Buddhism of the Vinayapiṭaka* (Delhi: Motilal Banarsidass, 1981).

13. Neihardt, *Black Elk Speaks,* p. 173.

14. Ibid., pp. 136–48.

15. Constantin Stanislavski, *Creating a Role,* translated by Elizabeth R. Hapgood (New York: Theatre Arts Books, 1961), p. 3 ff.

16. Ibid., p. 7.

17. See the forthcoming University of Chicago dissertation by Charles Hallisey, "Discovering Buddhist Devotion."

18. The status of Rūpa Gosvāmin and the other Vṛndāvana Gosvāmins is debated within the tradition. Some view them as saints who achieved perfection by means of *sādhana,* whereas others, as noted earlier (see the third section in Chapter 6), view them as incarnations of the eternally perfected.

19. Holt, *Discipline,* pp. 138–44.

Selected Bibliography

WORKS IN SANSKRIT, BENGALI, AND HINDI

Bharata, *Nāṭya-śāstra.* 2 vols. Edited with a Hindi translation by Madhusudan Shastri. Varanasi: Banaras Hindu University, 1971.

Gauragovindānanda Bhāgavatasvāmī. *Rāgānugā-bhakti-tattva-kusumāñjali.* Calcutta: Classic Press, 1948.

Gopālaguru Gosvāmin. *Gaura-Govindārcana-smaraṇa-paddhati.* In *Paddhati-trayam,* pp. 1–69. Edited by Haridāsa Dāsa. Navadvīpa: Haribol Kuṭīr, 1948.

——.*Śrī-Rādhā-Kṛṣṇayor Aṣṭa-kālīya-līlā-smaraṇa-krama-paddhati.* In *Paddhati-trayam,* pp. 70–88. Edited by Haridāsa Dāsa. Navadvīpa: Haribol Kuṭīr, 1948.

Haridāsa Dāsa. *Gauḍīya Vaiṣṇava Abhidhāna.* 3 vols. Navadvīpa: Haribol Kuṭīr, 1957.

Jānakiprasāda. *Nārāyana Bhaṭṭa-caritāmṛta.* Edited with a Hindi translation by Kṛṣṇadāsa Bābā. Mathurā: Kṛṣṇadāsa Bābā, 1955.

Jīva Gosvāmin. *Bhagavat Sandarbha.* Edited with an English introduction by Chinmayi Chatterjee. Calcutta: Jadavpur University, 1972.

——.*Bhakti Sandarbha.* Edited with a Bengali translation by Rādhāraman Gosvāmī Vedāntabhūṣan and Kṛṣṇagopāla Gosvāmī. Calcutta: University of Calcutta, 1962.

——. *Paramātma Sandarbha.* Edited with an English introduction by Chinmayi Chatterjee. Calcutta: Jadavpur University, 1972.

——.*Prīti Sandarbha.* Edited by Puridāsa. Vṛndāvana: Haridāsa Śarmaṇa, 1951.

Kavikarṇapūra. *Gaura-gaṇoddeśa-dīpikā.* In *Grantharatna-pañcakam.* Edited with a Hindi translation by Kṛṣṇadāsa Bābā. Mathurā: Kṛṣṇadāsa Bābā, Pusparāja Press, 1953.

Kṛṣṇadāsa Kavirāja. *Caitanya-caritāmṛta.* With the commentaries of Saccidānanda Bhaktivinod Thākur and Bārṣobhānabīdayita Dāsa. Calcutta: Gauḍīya Maṭh, 1958.

——.Govinda-līlāmṛta. 3 vols. With the commentary of Vṛndāvana Cakravartin. Edited with a Hindi translation by Haridāsa Śāstri. Vṛndāvana: Śrīgadādhara Gaurahari Press, 1977–81.

Kuñjabihārī Dāsa. *Mañjari-svarūpa-nirūpaṇa.* Rādhākuṇḍa: Ānanda Dāsa, 1975.

Mital, P. D. *Vraja kā Sanskṛitika Itihāsa.* Delhi: Rājakamal Prakāśan, 1966.

Narahari Cakravartin. *Bhaktiratnākara*. Edited by Navīnakṛṣṇa Paravidyālaṁkāra. Calcutta: Gauḍīya Maṭh, 1940.

Nareścandra Bansal. *Caitanya Sampradāya: Siddhānta aur Sāhitya*. Agra: Vinod Pustaka Mandir, 1980.

Niradprasād Nāth. *Narottama Dāsa o Tāhār Racanāvali*. Calcutta: University of Calcutta, 1975.

Rādhāgovinda Nāth. *Gauḍīya Vaiṣṇava Darśana*. 5 vols. Calcutta: Pracyabani Mandir, 1957–60.

Rāmānuja. *Śrībhāṣya*. Edited by Vasudeva Sastri Abhyankar. Bombay: Nirnayasagar Press, 1915.

Rupa Gosvamin. "Aṣṭa-kālīya-līlā-smaraṇa-maṅgala-stotraṃ." In *Early History of the Vaisnava Faith and Movement in Bengal*, pp. 673–75. Edited by S. K. De. Calcutta: Firma K. L Mukhopadyay, 1961.

______. *Bhaktirasāmṛtasindhu*. With the commentaries of Jīva Gosvāmin, Mukundadāsa Gosvāmin, and Viśvanātha Cakravartin. Edited with a Bengali translation by Haridāsa Dāsa. Navadvīpa: Haribol Kuṭīr, 1945.

______. *Bhaktirasāmṛtasindhu*. With the commentary of Jīva Gosvāmin. Edited by Gosvāmī Damodaraśāstri. Varanāsi: Acyuta Granthamālā Series, 1977

______. *Bhaktirasāmṛtasindhu*. With the commentaries of Jīva Gosvāmin and Viśvanātha Cakravartin. Edited with a Hindi translation by Śyāmadāsa. Vṛndāvana: Harinām Press, 1981.

______. *Bhaktirasāmṛtasindhu*. Vol. 1. Edited with an English translation and notes on the commentaries of Jīva Gosvámin, Mukundadāsa Gosvāmin, and Viśvanātha Cakravartin by Bon Mahārāja. Vrindaban: Institute of Oriental Philosophy, 1965.

______. *Ujjvalanīlamaṇi*. With the commentaries of Jīva Gosvāmin and Viśvanātha Cakravartin. Edited by Puridāsa. Vṛndāvana: Haridāsa Śarmaṇa, 1954.

______. *Upadeśāmṛta*. With several Sanskrit and Bengali commentaries. Edited with a Bengali translation by Bhaktikevala Auḍulomi Mahārāja. Calcutta: Śrī Gauḍīya Sampradāya, at Gauḍīya Maṭh, 1980.

Rūpa Kavirāja. *Rāgānugāvivṛtti*. Vrindaban Research Institute, manuscript #1194, Vṛndāvana, U.P., India. Also, edited with a Hindi translation by Haridāsa Śāstri. Vṛndāvana: Kṛṣṇadāsa Bábā, 1968.

______. *Sārasaṅgraha*. Edited with an English introduction by Krishnagopal Goswami. Calcutta: University of Calcutta, 1949.

Sanātana Gosvāmin. *Bṛhad-bhāgavatāmṛta*. Edited with a Hindi translation by Śyāma Dāsa. Vṛndāvana: Harinām Press, 1975.

Seṭha Govindadāsa. *Vraja aur Vraja-yātrā*. Delhi: Bharatīya Viśva Prakāśana, 1959.

Sundarānandadāsa Vidyāvinod. *Vaiṣṇava Siddhānte Śrīgurusvarūpa*. Calcutta: Karuṇā Dāsa, 1964.

Viśvanātha Cakravartin. *Bhaktirasāmṛtasindhu-bindu*. Edited with a Hindi translation by Śyāma Dāsa. Vṛndāvana: Harinām Press, 1977.

———. *Rahğavartmacandrikā.* Edited with a Bengali translation by Prāṇ Kiśor Gosvāmin. Howrah: Vinod Kiśor Gosvāmin, 1965.

Viśvanātha Kavirāja. *Sāhitya-darpaṇa.* Edited with a Hindi translation by Śālagrāma Śāstri. Delhi: Motilal Banarsidass, 1977.

Vopadeva. *Muktāphala.* With the commentary of Hemādri. Edited by Iśvara Chandra Śāstri and Haridāsa Vidyābagisa. Calcutta Oriental Series, no. 5. Calcutta: Baidya Nath Dutt, 1920.

WORKS IN ENGLISH

Anantharangachar, N. S. *The Philosophy of Sādhana in Viśiṣṭādvaita.* Mysore: University of Mysore, 1967.

Archer, W. G. *The Loves of Krishna in Indian Painting and Poetry.* New York: Grove Press, 1957.

Beals, Ralph C. "Religion and Identity." *Internationales Jahrbuch für Religionssoziologie* 11 (1978): 147–62.

Berger, Peter, and Luckmann, Thomas. "Arnold Gehlen and the Theory of Institutions." *Social Research* 32, no. 1 (1965): 110–15.

———. *The Social Construction of Reality.* New York: Doubleday & Co., 1966.

Beyer, Stephan. *The Cult of Tara.* Berkeley: University of California Press, 1973.

———. "Notes on the Vision Quest in Early Mahāyāna." In *Prajñāpāramitā and Related Systems: Studies in Honor of Edward Conze,* pp. 329–40. Edited by Lewis Lancaster and Luis Gomez. Berkeley: Berkeley Buddhist Studies Series, 1977.

Bhandarkar, R. G. *Vaiṣṇavism, Śaivism and Minor Religious Systems.* Varanasi: Indological Book House, 1965.

Bhaṭṭāchārya, Śivaprasād. "Bhoja's Rasa-Ideology and Its Influence on Bengal Rasa-Śāstra." *Journal of the Oriental Institute* (University of Baroda) 13, no. 2 (December 1963): 106–19.

Biddle, Bruce J., and Thomas, Erwin J., eds. *Role Theory: Concepts and Research.* New York: John Wiley & Sons, 1966.

Bloomfield, Maurice. "On the Art of Entering Another's Body: A Hindu Fiction Motif." *Proceedings of the American Philosophical Society* 56 (1917): 1–43.

Byrski, Christopher M. *Concepts of Ancient Indian Theatre.* New Delhi: Munshiram Manoharlal Publishers, 1973.

Carman, John B. *The Theology of Rāmānuja.* New Haven: Yale University Press, 1974.

Capps, Donald. "Sudén's Role-Taking Theory: The Case of John Henry Newman and His Mentors." *Journal for the Scientific Study of Religion* 21, no. 1 (March 1982): 58–70.

Chakravarti, Sudhindra C. *Philosophical Foundation of Bengal Vaisnavism.* Calcutta: Academic Publishers, 1969.

Coomaraswamy, Ananda K. "Līlā." *Journal of the American Oriental Society* 61 (1941): 98–101.

______. *Introduction to The Mirror of Gesture.* Translated by Ananda Coomaraswamy and Gopala Duggirala. Cambridge: Harvard University Press, 1917.

Cua, A. S. *Dimensions of Moral Creativity: Paradigms, Principles, and Ideals.* University Park: Pennsylvania State University Press, 1978.

Dasgupta, Shashibhusan. *Obscure Religious Cults.* 3rd ed. Calcutta: Firma KLM, 1946.

De, S. K. *The Early History of the Vaiṣṇava Faith and Movement in Bengal.* 2nd ed. Calcutta: Firma K. L. Mukhopadhyay, 1961.

______. *History of Sanskrit Poetics.* 2 vols. 2nd ed. Calcutta: K. L. Mukhopadhyay, 1960.

______. *Sanskrit Poetics as a Study of Aesthetics.* With notes by Edwin Gerow. Berkeley: University of California Press, 1963.

Delmonico, Neal. "Time Enough for Play: Religious Use of Time in Bengal Vaiṣṇavism." Paper presented at the Bengal Studies Conference, June 1982.

Deutsch, Eliot. *Advaita Vedánta.* Honolulu: University of Hawaii Press, 1973.

Dhayagude, Suresh. *Western and Indian Poetics: A Comparative Study.* Pune: Bhandarkar Oriental Research Institute, 1981.

Dimock, Edward C., Jr. "Hinduism and Islam in Medieval Bengal." In *Aspects of Bengali History and Society,* pp. 1–12. Edited by Rachel Van M. Baumer. Honolulu: University Press of Hawaii, 1975.

______. *The Place of the Hidden Moon.* Chicago: University of Chicago Press, 1966.

Dimock, Edward C., Jr., and Levertov, Denise. *In Praise of Krishna.* New York: Doubleday & Co., 1967.

Eck, Diana. *Darśan: Seeing the Divine Image in India.* Chambersburg, Pa.: Anima Publications, 1981.

Eliade, Mircea. *The Myth of the Eternal Return.* Princeton: Princeton University Press, 1954.

______. *Myth and Reality.* New York: Harper & Row, 1963.

______. *Patterns in Comparative Religion.* Translated by Rosemary Sheed. New York: Meridan Books, 1963.

______. *The Sacred and the Profane.* New York: Harcourt, Brace & World, 1959.

Evans, Robert D. "A Contribution to a Bibliography of Bengali Vaiṣṇavism." Unpublished paper, University of Chicago, December 12, 1980.

Farquhar, J. N. *An Outline of the Religious Literature of India.* Delhi: Motilal Banarsidass, 1920.

Frye, Northrop. *Anatomy of Criticism.* Princeton: Princeton University Press, 1971.

Geertz, Clifford. *The Interpretation of Cultures.* New York: Basic Books, 1973.

Gerow, Edwin. *A Glossary of Indian Figures of Speech.* The Hague: Mouton, 1971.

Gerow, Edwin. "Rasa as a Category of Literary Criticism." In *Sanskrit Drama in Performance,* pp. 226–57. Edited by Rachel Van M. Baumer and James R. Brandon. Honolulu: University Press of Hawaii, 1981.

______."The Rasa Theory of Abhinavagupta and its Application." In *Literatures of India,* pp. 216–17. Edited by Edward C. Dimock, Jr., et al. Chicago: University of Chicago, 1974.

Gerow, Edwin, and Aklujkar, Asok. "On Śānta Rasa in Sanskrit Poetics." *Journal of the American Oriental Society* 11, no. 1 (January–March 1972): 80–87.

Ghose, N. C. *Introduction to Sree Gouranga Lilamrtam.* Calcutta: Nitya Swarup Brahmachary, 1916.

Gilson, Ettienne. *The Mystical Theology of Saint Bernard.* Translated by A. H. C. Downes. New York: Sheed & Ward, 1940.

Goldman, Robert. "A City of the Heart: Epic Mathurā and Indian Imagination." *Journal of the American Oriental Society* 106, no. 3 (July–September 1986): 471–83.

Gonda, Jan, gen. ed. *A History of Indian Literature.* 10 vols. Wiesbaden: Otto Harrassowitz, 1973–. Vol. 5: *Indian Poetics,* by Edwin Gerow.

Goshal, S. N. *Studies in Divine Aesthetics.* Santiniketan: Visva-Bharati, 1974.

Gnoli, Raniero. *The Aesthetic Experience According to Abhinavagupta.* Rev. 2nd ed. Varanasi: Chowkhamba, 1968.

Grotowski, Jerzy. *Towards a Poor Theatre.* Holstebro: Odin Teatrets Forlag, 1968.

Growse, Fredrick S. *Mathurā: A District Memoir.* New Delhi: Asian Educational Services, 1882.

Guerric of Igny. *The Liturgical Sermons of Guerric of Igny.* Translated by the monks of Mount Saint Bernard Abbey. Spencer, Mich.: Cistercian Publications, 1970.

Hakeda, Yoshita S. *Kūkai: Major Works.* New York: Columbia University Press, 1972.

Hawley, John S. *At Play with Krishna: Pilgrimage Dramas from Brindavan.* Princeton: Princeton University Press, 1981.

Hawley, John S., and Wulff, Donna M., eds. *The Divine Consort: Rādhā and the Goddesses of India.* Berkeley: Graduate Theological Union, 1982.

Hein, Norvin. *The Miracle Plays of Mathurā.* New Haven: Yale University Press, 1972.

______."A Revolution in Kṛṣṇaism: The Cult of Gopāla." *History of Religions* 25, no. 4 (May 1986): 296–317.

Hiriyanna, Mysore. *Art Experience.* Mysore: Kavyalaya Publishers, 1954.

Holt, John C. *Discipline: The Canonical Buddhism of the Vinayapiṭaka.* Delhi: Motilal Banarsidass, 1981.

James, William. *Principles of Psychology.* 2 vols. New York: Henry Holt & Co., 1890.

______.*The Varieties of Religious Experience.* New York: Longman, Green, & Co., 1902.

Jaspers, Karl. *Socrates, Buddha, Confucius, and Jesus: The Four Paradigmatic Individuals*. Translated by Ralph Manheim. New York: Harvest Books, 1957.

Kapoor, O. B. L. *The Philosophy and Religion of Śrī Caitanya*. New Delhi: Munshriram Manoharlal Publishers, 1977.

Kazantzakis, Nikos. *The Greek Passion*. Translated by Jonathan Griffin. New York: Simon & Schuster, 1953.

Keith, A. Berriedale. *The Sanskrit Drama*. London: Oxford University Press, 1924.

Kennedy, Melville T. *The Chaitanya Movement*. Calcutta: Association Press, 1925.

Kinsley, David R. *The Divine Player: A Study of Kṛṣṇa Līlā*. Delhi: Motilal Banarsidass, 1979.

______. *The Sword and the Flute*. Berkeley: University of California Press, 1975.

Law, Narendra Nath. "Śrī Kṛṣṇa and Śrī Caitanya." 2 parts. *The Indian Historical Quarterly* 23, no. 4 (December 1947): 261–99; and 24, no. 1 (March 1948): 19–66.

Leclercq, Jean. *The Love of Learning and the Desire for God*. New York: Fordham University Press, 1961.

Lester, Robert G. "Aspects of the Vaiṣṇava Experience: Rāmānuja and Pilllai Lokācārya on Human Effort and Divine Grace." *Indian Philosophical Annual* 10 (1974–75): 1–10.

______. "Rāmānuja and Śrī-Vaiṣṇavism: The Concept of Prapatti or Śaraṇāgati." *History of Religions* 5, no. 2 (1966): 266–82.

______. *Rāmānuja on the Yoga*. Madras: Adyar Library and Research Centre, 1976.

Levy, Paul. *Buddhism: A Mystery Religion?* London: University of London, Athlone Press, 1957.

Majumdar, A. K. *Caitanya: His Life and Doctrine*. Bombay: Bharatiya Vidya Bhavan, 1969.

Majumdar, R. C. *History of Mediaeval Bengal*. Calcutta: G. Bharadwaj & Co., 1973.

Malinowski, Bronislaw. *Magic, Science, and Religion*. New York: Doubleday & Co., 1948.

Martz, Louis L. *The Poetry of Meditation*. New Haven: Yale University Press, 1962.

Masson, J. L. "The Childhood of Kṛṣṇa: Some Psychoanalytic Observations." *Journal of the American Oriental Society* 94, no. 4 (1974): 454–59.

Masson, J. L., and Patwardhan, M. V. *Aesthetic Rapture*. Poona: Deccan College, 1970.

______. *Śāntarasa and Abhinavagupta's Philosophy of Aesthetics*. Poona: Bhandarkar Oriental Research Institute, 1969.

Moore, Sonia. *The Stanislavski System*. New York: Penguin Books, 1965.

Mulji, Karsandas. *History of the Sect of Maharajas, or Vallabhacaryas, in Western India*. London: Trubner & Co., 1865.

Neihardt, John G. *Black Elk Speaks.* New York: Pocket Books, [1932] 1972.

O'Connell, Joseph T. "Rāgavartmacandrikā of Viśvanātha Cakravartin." In *A Corpus of Indian Studies,* pp. 185–209. Edited by A. L. Basham et al. Calcutta: Sanskrit Pustak Bhandar, 1980.

______."Social Implications of the Gauḍīya Vaiṣṇava Movement." Ph.D. dissertation, Harvard University, 1970.

O'Flaherty, Wendy D. *Dreams, Illusion and Other Realities.* Chicago: University of Chicago Press, 1984.

______.*Women, Androgynes and Other Mythical Beasts.* Chicago: University of Chicago Press, 1980.

Otto, Rudolf. *Christianity and the Indian Religion of Grace.* Madras: Christian Literature Society for India, 1929.

______.*The Idea of the Holy.* Translated by John W. Harvey. London: Oxford University Press, 1923.

Pandey, K. C. *Abhinavagupta: An Historical and Philosophical Study.* Varanasi: Chowkhamba Sanskrit Series, 1963.

Pernet, Henry. " 'Primitive' Ritual Masks in the History of Religions." Ph.D. dissertation, University of Chicago, 1979.

Potter, Karl H. *Presuppositions of India's Philosophies.* Westport, Conn.: Greenwood Press, 1963.

Raghavan, V. *Bhoja's Śṛṅgāra Prakāśa.* Rev. 3rd ed. Madras: Punarvasu, 1978.

______.*The Number of Rasa-s.* Madras: Adhar Library and Research Centre, 1967.

Ramanujan, A. K. *Hymns for the Drowning.* Princeton: Princeton University Press, 1981.

______.*Speaking of Śiva.* Baltimore: Penguin Books, 1973.

Sarbin, Theodore R. "Role Theory." In *Handbook of Social Psychology,* pp. 223–58. Edited by Gardner Lindsey. Cambridge: Addison-Wesley, 1954.

Schutz, Alfred. *Collected Papers I: The Problem of Social Reality.* Edited by Maurice Natanson. The Hague: Martins Nijhoff, 1973.

Sekaquaptewa, Emory. "Hopi Indian Ceremonies." In *Seeing with a Native Eye,* pp. 35–43. Edited by Walter Capps. New York: Harper & Row, 1976.

Sen, Dinesh Chandra. *Chaitanya and His Age.* Calcutta: University of Calcutta, 1924.

Siegal, Lee. *Sacred and Profane Dimensions of Love in Indian Traditions as Exemplified in the Gitagovinda of Jayadeva.* Delhi: Oxford University Press, 1978.

Singer, Milton, gen. ed. *Krishna: Myths, Rites, and Attitudes.* Chicago: University of Chicago Press, 1968.

Simonov, P. V. "The Method of K. S. Stanislavski and the Physiology of Emotion." In *Stanislavski Today,* pp. 34–43. Edited by Sonia Moore. New York: American Center for Stanislavski Theatre Art, 1973.

Smith, Jonathan Z. "The Bare Facts of Ritual." *History of Religions* 20, no. 1 & 2 (August–November 1980): 112–27.

Soderblom, Nathan. *The Living God: Basal Forms of Personal Religion*. London: Oxford University Press, 1931.

Spiro, Melford E. "Religion: Problems of Definition and Explanation." In *Anthropological Approaches to the Study of Religion*, pp. 85–126. Edited by Michael Banton. London: Tavistock Publications, 1966.

Stanislavski, Constantin. *An Actor Prepares*. Translated by Elizabeth R. Hapgood. New York: Theatre Arts, 1946.

______. *Creating a Role*. Translated by Elizabeth R. Hapgood. New York: Theatre Arts, 1961.

Stewart, Tony K. "The Biographical Images of Kṛṣṇa-Caitanya: A Study in the Perception of Divinity." Ph.D. dissertation, University of Chicago, 1985.

Sudén, Hjalmar. "What Is the Next Step to Be Taken in the Study of Religious Life." *Harvard Theological Review* 58 (1965): 445–51.

Tagore, Rabindranath. *Sādhana*. Madras: Macmillan, 1913.

Turner, Victor. *Dramas, Fields, and Metaphors*. Ithaca: Cornell University Press, 1974.

Underhill, Evelyn. *Mysticism*. New York: Noonday Press, 1955.

Van Buitenen, J. A. B. *Rāmānuja on the Bhagavadgītā*. Delhi: Motilal Banarsidass, 1968.

Vaudeville, Charlotte. "Braj, Lost and Found." *Indo-Iranian* Journal 18 (1976): 195–213.

Wach, Joachim. *The Comparative Study of Religion*. Edited by Joseph M. Kitagawa. New York: Columbia University Press, 1958.

Waghorne, Joanne Punzo, and Cutler, Norman, eds., *Gods of Flesh, Gods of Stone*. Chambersburg, Pa.: Anima Publications, 1985.

Walimbe, Y. S. *Abhinavagupta on Indian Aesthetics*. Delhi: Ajanta Publications, 1980.

Wallace, Anthony F. C. *Religion: An Anthropological View*. New York: Random House, 1966.

Weigle, Martha. *Brothers of Light, Brothers of Blood: The Penitentes of the Southwest*. Albuquerque: University of New Mexico, 1976.

Wiles, Thomas J. *The Theater Event*. Chicago: University of Chicago Press, 1980.

Wilson, H. H. *The Religious Sects of the Hindus*. Calcutta: Susil Gupta, [1861], 1958.

Wulff, Donna M. *Drama as a Mode of Religious Realization: The Vidagdhamādhava of Rūpa Gosvāmin*. Chico, Calif.: Scholars Press, 1984.

Index